The Theoretical Evolution
of International Political Economy

The Theoretical Evolution
of International Political
Economy

_____ *A Reader*

Edited by

GEORGE T. CRANE

ABLA AMAWI

New York   Oxford
Oxford University Press
1991

Oxford University Press

Oxford   New York   Toronto
Delhi   Bombay   Calcutta   Madras   Karachi
Petaling Jaya   Singapore   Hong Kong   Tokyo
Nairobi   Dar es Salaam   Cape Town
Melbourne   Auckland

and associated companies in
Berlin   Ibadan

Published by Oxford University Press, Inc.,
200 Madison Avenue, New York, New York 10016

Library of Congress Cataloging-in-Publication Data
The Theoretical evolution of international political economy:
a reader / edited by George T. Crane, Abla Amawi.
p. cm.   ISBN 0-19-506012-1
1. International economic relations.
I. Crane, George T., 1957–. II. Amawi, Abla.
HF1359.T44   1991   337—dc20   90-35879

9 8 7 6 5 4

Printed in the United States of America
on acid-free paper

Preface

The idea for this volume came in the midst of a frustrating search for course material on theories of international political economy (IPE). It was painfully clear that key readings were difficult to marshal for classroom use. This collection is designed as a remedy, making a number of important works of IPE theory readily available. The framework is familiar: the tripartite categorization of liberalism, realism, and Marxism popularized by, among others, Robert Gilpin. As a commentary on the continuing theoretical evolution of IPE, however, this book also includes more recent arguments that do not fit neatly into established categories. The choice of readings attempts to balance theoretical significance with pedagogical utility. Thus, as with any project of this sort, not every major work is included. We hope our selections will prove helpful to most teachers and students of IPE.

This reader would have been impossible without the outstanding contribution of the coeditor, Abla Amawi. While I was away teaching for a year in China during the 1988–1989 academic year, Abla handled all administrative matters, which were considerable, as well as preliminary editorial work. She also provided excellent suggestions for the introduction and participated in choosing a number of readings. The volume benefitted in myriad ways from her extraordinary talent.

Many others helped us complete this project. The introduction and reading selections were improved by the comments and criticisms of Robert Gilpin, James Caparoso, Larry Malone, Kate Manzo, and two anonymous readers from Oxford University Press. Students at the University of Wisconsin-Madison, Georgetown University, the Johns Hopkins-Nanjing University Center for Chinese and American Studies, and Williams College have been unwitting contributors. They provided the classroom testing ground for these and many other chapters and articles. While I was in China, the staff of the Washington Office of the Hopkins-Nanjing Center facilitated communications with the U.S., difficult in the best of times and complicated further by the events of spring 1989.

We also wish to thank Reem Abu-Lughod for typing the manuscript. Ayda Amawi and William Cordes helped at crucial moments. Our respective spouses, Michael Fischbach and Maureen Strype, encouraged and supported us.

The Chinese students I had the pleasure of teaching in 1988–1989 warrant a special note of thanks. Their remarkable effort to spark peaceful political change in their country, against aweful repression, provides a stirring lesson in courage and conviction. They embody the power of ideas, in this case the intertwined concerns of democracy and development. When they finally succeed, they may well inspire new thinking on the political economic forces that they contend with daily.

Williamstown, Mass.   G. T. C.
June 1990

Contents

The Theoretical Evolution
of International Political Economy

Theories of International Political Economy

The purpose of this collection is to outline the major theoretical currents of international political economy. The approach is historical because core assumptions of key schools of thought were posited and debated long ago. Moreover, past scholarship continues to shape the present. The volume offers representative examples of three classical paradigms—mercantilism, liberalism, and Marxism—to illustrate the early conceptual contours of the field. These are complemented by modern interpretations of the major theories to demonstrate the evolution of each. More recent arguments, however, move beyond established theoretical boundaries. This collection therefore includes contemporary approaches that cut across the major schools of thought. But before each theory is briefly introduced, some basic definitional questions must be addressed.

What Is International Political Economy?

What is the common ground of the various theories of international political economy (IPE)? The most direct answer to this question is found in the name itself. _Political economy_ suggests a focus on phenomena that lie at the crossroads of the traditional fields of political science and economics. It seeks to explain how political power shapes economic outcomes and how economic forces constrain political action. Robert Keohane provides more depth to this definition:

> We can view international political economy as the intersection of the substantive area studied by economics—production and exchange of marketable

means of want satisfaction—with the process by which power is exercised that is central to politics.[1]

Although divergent beliefs are held as to the direction and strength of the relationship of politics and economics, investigating their connection is the stuff of political economy. But political economy is not simply an amalgam of the two traditional fields; rather, it attempts a new synthesis. Implicit in the endeavor of political economy is a critique of the scope and methods of both economics and political science.[2] To the political economist, ceteris paribus assumptions and numerous "exogenous" variables rob neoclassical economics of its explanatory power. Likewise, much of political science pays insufficient attention to how economic processes and structures might influence the play of power. By contrast, a theory of political economy should be judged precisely by how well it explains the interaction of politics and economics.

International political economy primarily, though by no means exclusively, concentrates upon activities taking place *among* international actors: states, global corporations, international organizations, and so on. Such a focus naturally impinges upon the customary domain of international relations. Indeed, the recent renaissance of IPE is in part a response to perceived shortcomings in the dominant realist paradigm of international relations. In the postwar era, realism has tended to separate politics from economics and define power largely in military–political terms.[3] As Susan Strange argues, this limits understanding of "structural power," in which international structures of production play an important part.[4]

Moreover, unlike Kenneth Waltz's neorealism, IPE ranges well beyond the international level of analysis.[5] Although the lion's share of IPE work is taken up by a few internationally oriented empirical topics (trade, finance, investment, and development), a significant portion of the field looks *within* particular states and national markets to understand how domestic forces influence international action. The boundary between the international and the intranational is not rigidly fixed in the IPE literature. Some IPE arguments are applied across levels of analysis, while others logically lead to a national or subnational focus.

The task of defining IPE, therefore, is not an easy one. The lines between politics and economics, as well as those between international and domestic, are blurred. Moreover, the variety of theoretical frameworks in IPE creates even greater difficulty in delineating the field of study, for each defines the field differently. A number of categorizations for the various theories have been put forth, but arguably the most fully developed, and most influential, is Robert Gilpin's.[6] He points to three major schools of thought: "nationalist" (also referred to as mercantilist, realist, or statist), "liberal," and "Marxist." Each possesses a distinct universe of discourse with a long intellectual tradition as well as a variety of contemporary manifestations. Gilpin's typology is useful as a heuristic device for understanding the evolution of IPE, because it encompasses many of the key theoretical controversies. He concisely captures the diversity of the field.

   Gilpin's framework, however, has two shortcomings. First, he refers to the three schools of thought as "ideologies" because of the tenacity with which adherents maintain intellectual commitments, even in the face of contrary logic or evidence.[7] This suggests a powerful aversion to theoretical change. However, as this volume attempts to demonstrate, significant revision has occurred in IPE. Much of this is due to criticisms generated within one school of thought or another. Reappraisals of imperialism and dependency theory have, by and large, been the result of debates among Marxist scholars. On the other hand, interparadigmatic dialogue, for all of its inherent difficulties, has also led to theoretical revision. For example, a lively debate has raged between liberal proponents of interdependence and neomercantilists, an exchange that has brought greater sophistication to each.

   Furthermore, ideology implies that adoption of a given perspective entails acceptance of a set of empirical *and* normative propositions. This, however, is not even true for Gilpin himself. He prefers liberal prescriptions while recognizing the validity of certain nationalist and Marxist arguments.[8] Such eclecticism is characteristic of other scholars who do not necessarily agree with all the normative implications of a particular empirical theory. Although it is incorrect to argue that IPE is objective and free of ideology, it would be equally inaccurate to err in the opposite direction and assume that all who work in the field hold onto consistent and integrated normative and empirical beliefs.

   Gilpin's framework faces a second problem. The boundaries of the three major perspectives have blurred somewhat in recent years. Rational choice analysis, for example, is an offshoot of liberalism that has been integrated into statist and some Marxist arguments. Regime theory and hegemonic stability theory likewise cut across theoretical divides. These sorts of arguments transcend the tripartite categorization of IPE theories. Contemporary developments, however, have historical roots. Thus, a discussion of the *continuing* theoretical evolution of IPE, and how new ideas have outgrown established categories, must begin with classical thinkers.

The Rise and Fall of Mercantilism

The earliest systematic theorizing on IPE is the classical mercantilist thought of the sixteenth through nineteenth centuries. In the words of Jacob Viner, mercantilism is: ". . . a doctrine of extensive state regulation of economic activity in the interest of the national economy."[9] Eli Heckscher makes a similar point: "Mercantilism would . . . have had all economic activity subservient to the state's interest in power."[10] Drawing on the political realism of Thucydides, Machiavelli, and Hobbes, mercantilists argue that if formal authority does not constrain the pursuit of self-interest, the result is likely to be a brutal "state of nature." Mercantilist thinkers thus believe in the necessity of public authority to translate individual interests into universal good.[11] Their analysis tends to focus on collectivities instead of individuals; maximizing *state* power and wealth is the best means of ensuring *public* welfare. Raison d'état is a central economic principle, and state interest does, and should, determine economics in mercantilist theory.

For early mercantilists, wealth inheres in the absolute amount of gold or silver held in the public treasury. Since these are tangible assets of state power, the government intervenes in the economy both domestically and internationally to maximize holdings of specie. Domestically, extensive intervention takes the form of national economic consolidation and more effective collection of revenues. Internationally, intervention leads to protectionism, the lasting contribution of mercantilism to the lexicon of IPE. A state's balance of trade is considered a central element of the international balance of power. The country must run a balance of trade suprlus, maintaining an inflow of specie, to support its position in the international system of self-interested states. Insofar as mercantilists see this world political–economic struggle as a zero–sum game, no effort should be spared in protecting the national economy.

Mercantilism served for decades as a way of looking at the world and as a guide for policy. Its acceptance made sense politically in that it evolved simultaneously with the rise of the modern state. State-building absolutist princes encouraged and embraced mercantilist thought. By the late eighteenth century, however, it began to run into vexing anomalies. Adam Smith posited a withering critique of mercantilism, arguing that state interests will not be maximized if prescriptions of intervention and protectionism are followed. The points raised by Smith are not the only difficulties mercantilism encounters. The international economic dominance of the United Kingdom, the avowed champion of liberalism, strengthen the free trade argument by giving it a political foundation. While this may have merely been an ideological façade to hide the economic power of the hegemonic British state, the liberal explanation of *Pax Britannica* is sufficiently powerful to undercut mercantilist reasoning.

Mercantilism, however, does not die. Friedrich List recasts it in somewhat diluted form. List is very pointed in his critique of liberalism, but his revision of mercantilism ultimately accepts the logic of free trade. His infant industry argument is basically ad hoc, explaining why certain states, Germany in particular, should be given a chance, through protectionism, to ascend the ranks to industrialized status. While his fervent nationalism maintains the spirit of mercantilism, he is unable to defeat the powerful logic of comparative advantage and free trade. The German Historical School also carries on mercantilist assumptions in the aftermath of List's compromise. Gustav Schmoller, the preeminent representative of the Historical School, reasserts the nationalist premise that economic processes are fundamentally shaped by political forces.[12] He rejects the liberal emphasis on the individual, as does List, in favor of a focus on the state and other key sociopolitical institutions. Schmoller's own work suggests an interactive relationship of politics and economics, but some of his colleagues propose unabashedly politicist arguments.[13]

The influence of the Historical School is, however, short-lived. Its inability to develop a distinct and viable approach to IPE is rooted in its methodology.

Schmoller has an aversion to abstract grand theory, which separates him from both liberals and Marxists. Instead, he emphasizes historical monograph, a method that robs the perspective of its potential to evolve into a mature theoretical system. For the most part, the Historical School does not have core generalizations; it is more the shared intellectual penchant of a number of careful scholars. Thus, while economic nationalism lived on in the policies of countries such as Germany and the United States, and flourished during the political–economic crisis of the 1930s, the theoretical ground of mercantilism was undermined by liberalism. It would take many decades until it was revised into a major IPE theory.

Smith, Ricardo, and the Origins of Liberal IPE

Adam Smith does much more than criticize mercantilism; he posits an alternative paradigm.[14] At the core of his system is an assumption, made by mercantilists as well, of rational egoism. But for Smith this does not imply a Hobbesian state of nature. Instead, the "invisible hand" of the market naturally ensures that the pursuit of self-interest, in and of itself, will lead to the public good. Freedom is a crucial element in this process. Freely functioning markets, based upon a division of labor, will ultimately maximize efficiency and prosperity. Smith recognizes that "irrational" political intervention or economic collusion, can, and in his time did, undermine the felicitious consequences of free markets. His arguments for freer trade and limited government intervention are tempered by an understanding of political realities. Nonetheless, Smith clearly believes that economics has a logic distinct from politics and should not be unduly hampered by political machinations. He turns mercantilism on its head.

David Ricardo adds to both the content and the methodology of Smith's liberalism. In Smith's argument, gains from foreign trade are largely based on absolute advantage. Free trade is universally beneficial when each nation can produce some particular commodity more efficiently than any other. Ricardo takes the defense of free trade a step further. Even if a country has no absolute advantage, it can still derive gains from trade. His famous example of England and Portugal demonstrates how trade based on comparative costs necessarily leads to mutual economic advantages. Ricardo thus develops a point that Smith raises but does not fully explicate: specialization and free trade based on comparative advantage.

Ricardo also revises the methodological ground of Smith's liberalism. Smith is philosophical and holistic, placing political economy within a moral and historical context. Ricardo, on the other hand, is narrower in his analytic focus, but more rigorous.[15] He separates economic questions from political and sociological issues and generates more parsimonious explanations. Although Smith's insights into the functioning of free markets places individual economic actors in the center of his analysis, Ricardo lays the foundation for the methodological individualism that is a hallmark of liberal IPE. By the end of the nineteenth century, liberal analysis conceptualizes political economy almost wholly in terms of interrelationships among rational individuals. This approach is applied across levels of analysis: individual consumers in a national economy; individual groups; and individual national economies interacting with one another. Ricardo's method, while not as formal and abstract as later economists, foreshadows a distinctly liberal methodology.[16]

Smith and Ricardo also draw out the implications of comparative advantage and free trade. As market mechanisms are carried from the national economy to the world economy, a certain equilibrium will be established in international economic relations. The quantity theory of money (originally mercantilist but refined by liberals), directly ties changes in reserves to changes in domestic prices. This ensures that trade and payments imbalances are automatically brought back into balance. A freely functioning international market should be self-corrective. Additionally, if all countries specialize and trade, then economic prosperity is diffused throughout the world. Any country will therefore find a niche and benefit if liberal principles are conscientiously followed. International equilibrium and growth should arise from comparative advantage and free trade.

The economic focus of classical liberalism raises a question of whether this is truly a theory of IPE (i.e., of both politics and economics) or whether it subordinates politics to the point of nonexistence. Closer inspection suggests that liberalism does embrace a theory of politics. It views politics as the rational management of a naturally harmonious community. This shapes the liberal notion of war and international relations. Nineteenth-century liberals argue that war is: ". . . the natural state of men ignorant of the laws of political economy."[17] In other words, if free trade were encouraged, the likelihood of political conflict and war diminishes. Because war undermines productive capacity and saps national wealth and power, peace is logically in the interest of every state. War is not an outgrowth of conflicting national interests, but it arises from "national interests ill understood." Political economy is "the science *par excellence* of peace." For early liberals, free trade is so powerful a mutual interest that international political institutions are not necessary to ensure world peace. IPE thus contemplates the relationship of economics and politics, albeit in a laissez-faire fashion.

Liberal arguments changed the face of IPE as a field of study. Modern economics was professionalized on the basis of liberal tenets and texts. In academia and government alike, liberal IPE had many faithful defenders by the turn of the century. What explains the vitality of this approach? Liberal free trade and comparative advantage arguments not only explained new "facts" of international economic life, but they also seemed to corroborate the economic success of Great Britain. Although most countries clung to protectionism, liberalism predicted ultimate failure of this course. And those predictions enjoyed credence insofar as the "beggar-thy-neighbor" trade conflicts in the 1920s and 1930s contributed to the Great Depression. Liberalism thus offered powerful and parsimonious answers to the central questions of IPE.

Additionally, prior to World War I, liberal notions of politics were very alluring, more for their logical coherence and rationalist optimism, than correspondence with reality. In this period, liberalism influenced the rise of idealism in international relations. Norman Angell's argument on the futility and economic waste of war is a most illustrative example.[18] After World War I, Wilsonian idealism departed from the liberal disdain of international organizations, but continued to accept the key principle of fundamentally harmonious interests among states.[19] Thus, it would not be until after the Great Depression and World War II that liberal IPE would be faced with the need of revision in light of Keynesian theory. This is not to say that the early liberals went completely unchallenged. Besides the continuing practical application of protectionism (it speaks to the politician), Marxism presented a direct assault on liberalism.

Marx and the Early Marxists

One of the most problematic aspects of Marxist IPE is the dearth of Karl Marx's own writings on the international realm. His thoughts on IPE, found mostly in newspaper articles from the 1850s, are scattered and not systematically developed.[20] The evolution of this perspective relies heavily on those who have come after Marx. The core of this approach is, nonetheless, of Marx's own making. Unlike mercantilism and liberalism, Marx does not assume that people are immutably rational egoists. Human motivations and orientations are largely shaped by the material environment of a particular time. This is the foundation of one of Marx's most basic premises: historical materialism. Marxists hold that the way society is organized to produce its economic necessities essentially conditions political activity as well as individual consciousness. Marxist analysis generally interprets the relationship between economics and politics as being dominated by economic forces expressed in a historically specific mode of production.

For Marx, history is marked by epochal changes in modes of production (e.g., feudalism to capitalism); that is, historical materialism is a theory of history. Although he contemplates a vast sweep of historical change, Marx focuses on defining the "laws of motion" of the capitalist mode of production. Capitalism is marked by a distinct social organization: a class structure of owners of the means of production pitted against workers possessing only the capacity to work—labor power. This distinction is based on both alienation (of workers from the products of their labor) and exploitation (of workers by capitalists). Class structure thus provides the necessary social prerequisites for the accumulation of capital.

To simplify a complex argument, Marx holds that the surplus value capitalists reap by selling the commodities produced by workers must be productively reinvested. If a capitalist fails to accumulate, other direct competitors will drive him from the market. In essence, accumulation is the driving force of capitalism; it is the primary source of unprecedented material wealth and technological innovation under conditions of capitalism.[21] But accumulation is not sufficient to guarantee the permanence of capitalism. Rather, crisis, break-down, and revolution are an inherent part of this mode of production. Gener-ally, Marx argues that an intrinsic contradiction, the growing inequality of the two classes, will eventually goad the working class to foment revolution and usher in a new mode of production.

Marx himself derives very few empirically testable IPE arguments from these core assertions. Indeed, the concept of "testing" itself has no place in Marxian epistemology. Marx's methodology involves discovering historically specific manifestations of core truths, not the linear causation of antecedent to consequent. For example, he believes the economic power (i.e., the ability to accumulate) of the capitalist mode of production will change the world. Other modes of production will be subsumed by the social and productive relations of capital.[22] No specific theoretical explanation answers why capital must expand globally. But the logic of historical materialism suggests that all countries experi-ence a stage of capitalist development: "The country that is more developed industrially only shows, to the less developed, the image of its own future."[23] In the later years of his life, there is some evidence that Marx began to revise his ideas. The "late Marx" suggests that the transition from one mode of produc-tion to another is more complex than suggested in his earlier writing.[24] The hints of revision, however, did not significantly influence those after Marx who constructed more thoroughgoing IPE arguments.

This enterprise is taken up by, among others, Rosa Luxemburg, Rudolf Hilferding, Nikolai Bukharin, and V.I. Lenin.[25] Each makes a specific contri-bution to subsequent Marxist IPE theory. Although often at odds with one another, enough common ground exists for synthesis, which Lenin provides. For Lenin, the central issue of IPE is imperialism, a stage of capitalism which involves, among other things, the expansion of capital internationally. He explains both the causes and the consequences of imperialism, albeit in dialec-tical fashion. Borrowing heavily from Hilferding, Lenin argues that the inter-nal dynamics of capital result in global expansion. He contends that competi-tive accumulation, and the resultant concentration of capital, lead to larger and larger national monopolies. As opportunities for continued growth and accumulation shrink domestically, these monopolies cast an eye to foreign lands. This process takes a number of different forms: the search for new investment opportunities; the need for new markets; or the demand for raw materials.

The existence of territorial states, however, complicates the international-ization of capital. The power of monopoly capitalists allows them to control their respective state apparati; state power is an instrument of class domina-tion. As competitive accumulation is carried to the international arena, mili-tary force thus becomes a means of securing economic gain. At first, military power is used against the targets of imperialism in the "pacification" of colo-nies. Eventually, the logic of accumulation leads national monoplies to com-pete directly against one another, and interimperialist war ensues. Lenin's notion of imperialism sees world war as the ultimate political outcome. Short of Armageddon, Lenin also suggests other effects of imperialism. He argues that as capital seeks new foreign outlets in the periphery of the world econ-

omy, these areas experience accelerated capitalist development.[26] He reinforces Marx's expectation that capitalism will spread to all corners of the
world. Moreover, Lenin states that as the centers of industrial vitality shift
toward the periphery, the traditional metropoles of capitalism experience
economic decay. Lenin therefore suggests that Europe must face a dilemma:
world war or economic decline.

Marx and the early Marxists add a new dimension to IPE. At a moment
when liberalism, in the thrall of Ricardo, is increasingly focused on economics
alone, Marx broadens the scope of analysis in an epic theory of *political*
economy. Although economic forces are given pride of place, political implications are very much a part of Marxist analysis. Furthermore, historical materialism is a theory of economic, political, and social change, promising an
explanatory dynamism that liberalism lacks. It is also a rationale for revolution that draws the attention of political activists. Thus, while Marxism had a
miminal impact on turn-of-the-century economists, it shifts the problematique
and creates a new audience, a new political economy.[27]

In sum, by the early twentieth century, three major theories dot the IPE
landscape. The mercantilist tradition, at this time, is in decline. Its post-
Smithian revisions are largely ad hoc and its explanatory power is weak.
While protectionism continues, mercantilist theory has waned. Liberals and
Marxists alike interpret the state, the central mercantiist unit of analysis, as an
epiphenomenon. On the other hand, the grand theory of Marxism and the
parsimonious rigor of liberalism occupy the center of the intellectual debate.
Changes in international circumstances in the post-World War II period
would, however, profoundly influence all three theories.

Modern Revisions of Liberal IPE

A number of notable additions have been made to liberalism since the turn of
the century. The Marginalist problem shift is possibly the most important in
the pre-World War I era, contributing both substance and methodology. Marginalist neoclassical economics, however, does not fundamentally change the
principles of liberal IPE as developed by Ricardo. The later Heckscher-Ohlin
factor-proportions theorum is a refinement, not a thoroughgoing revision, of
the Ricardian tradition. The single greatest modern challenge to liberal notions of IPE before World War II, from within its own ranks, is that raised by
John Maynard Keynes.

The Impact of Keynes on IPE

Quite simply, Keynes changed the language of liberal economic thought. The
thrust of his complex and controversial argument is aimed at the liberal belief
that markets inherently tend towards a socially beneficial equilibrium.[28] Instead, he contends that production and consumption more often than not
balance where unemployment remains at a relatively high level. Keynes' re-

sponse to this problem breaks with liberal precedent. He advocates state intervention in the economy to stimulate both employment and investment. Although such action, in Keynes's mind, should be countercyclical and complementary to market forces, this point challenges the basic liberal argument that economic prosperity can best be gained without state management. Additionally, the Keynesian focus on aggregates, in place of methodological individualism, questions the traditional unit of liberal analysis.[29]

In judging Keynes's contribution, it must be kept in mind that his work is a product of its theoretical environment. Keynes was writing when the analytic separation of politics and economics was the norm of his profession. He speaks directly to economists in the professional sense; his aim is to contribute to economic science, not political economy. Whether intentional or not, however, Keynes does add to IPE theory.

Generally, Keynes argues that a country must have full employment as its primary economic goal to avoid debilitating and unnecessary recessions. This priority both influences and is influenced by international economic policies. To secure full employment, for example, a country will want to run a balance of trade surplus, since exports contribute to effective demand. By the same logic, every country will want a trade surplus. This implies that protectionism may follow from domestic Keynesian policies. Although Keynes exhibits a nationalist streak, he is not a protectionist.[30] He understands that devastating beggar-thy-neighbor policies of the interwar years must be avoided. This is best accomplished through international economic cooperation, an end he ardently worked toward throughout his life. He actively lobbied against the Gold Standard, which forced domestic deflation and raised unemployment. And he supported efforts to create international institutions to provide national leaders with some flexibility in balancing domestic priorities with international obligations.

Keynes shifts the debate in liberal IPE away from the issue of whether international intervention should occur, to what sort of institutions should be created. The key is his basic acceptance of economic management, and consequently a role for the state at the international level. As with his domestic interventionist ideas, such management should complement, not contradict, underlying world market forces. Keynes thus stays within the purview of liberal thought.

However, by bringing the state into his analysis, Keynes poses new problems for liberalism. The issue of *how* states should intervene internationally eventually leads to the question of *why* states actually intervene as they do. A series of events in the late 1960s and early 1970s raised this issue. The Bretton Woods system disintegrated, calling into question liberal arguments regarding international cooperation. OPEC's successful manipulation of international oil markets could not be adequately explained by liberal assumptions alone; the market was not free, competition was far from perfect. In addition, many less developed countries were not gaining the benefits of comparative advantage, challenging liberalism's predicted international diffusion of economic development. These events suggest that markets simply behave differently than liberal theory holds; in a word, they appear to be politicized. The key anomaly facing liberal IPE today is explaining how politics shapes economic forces—the converse of the traditional expectation that economics should determine politics. Increasingly, the theoretical progress of liberalism can be gauged by how well its assumptions explain the political dynamics of international markets.

The Tenacity of Economic Orthodoxy

A second strand of modern liberalism ignores the anomalies at hand and retreats to neoclassical orthodoxy. Although these arguments address relevant subject matter, they are not really theories of IPE as such; they do not question the relationship between politics and economics. The neoclassical approach, true to its Marginalist roots, maintains the analytic separation of the two. Persistent and economically inefficient political intervention in the market is not explained; rather, it is dismissed as "irrational." This economistic impulse is hardly new; it has been extant since at least the Marginalist "revolution" and was buttressed against the onslaught of Keynesianism. One of the most pointed orthodox critiques of Keynesian IPE was Milton Friedman's argument against the fixed exchange rate system of Bretton Woods.[31] He argued that the self-regulating, self-correcting market dynamic would yield superior economic outcomes compared to interventionist policies. His position was based on a theory of economics alone; there was little consideration of why markets might in fact be politicized.

Neoclassical economic orthodoxy is not a genuine advance in liberal IPE. Rather, it is a retreat to the original arguments that were so seriously challenged by the international economic events of the late 1960s and early 1970s. Instead of providing "new facts," it merely repackages the old. This regressive form of liberalism places greater value on internal logical coherence than on external correspondence with empirical evidence; it is simply impermeable to theoretical challenges of linking politics and economics.[32]

The Interdependence Research Program

Analyzing economics and politics is the theoretical project of the interdependence approach. The term *interdependence* was coined in the late 1960s to describe the growing interconnectedness of national economies. Deepening trade, investment, and monetary ties among advanced industrial countries draw national economies closer together and create a truly global economy. The overall effect is debilitating to national economic autonomy. Each country is faced with a wider array of disrupting economic forces, but each has less capacity to control such disturbances. Interdependence thus creates a paradox; it produces incentives for both cooperation and competition. Successful management of transnational economic relations may best be achieved through policy coordination and cooperation. But frustrated state leaders could also attempt to regain national economic control by means of damaging beggar-thy-neighbor policies. Interdependence analysis focuses on this tension of state and world market.

Some liberal analysts see the contradiction of interdependence ultimately resolving itself in favor of economic forces. States, they hold, must give way to markets. Perhaps the most daring propositions come from those liberals who analyze the rise of transnational corporations. State sovereignty is held "at bay" by the power of these gigantic companies.[33] National political structures, rendered increasingly obsolete by global economic forces, will have to surrender a measure of their authority to transnational organizations better suited to respond to the new international environment. These are eminently liberal expectations, consistent with the belief that economic logic should not be thwarted by "irrational" politics. Indeed, such optimistic interdependence liberals go further than their classical brethren, arguing that technological and market changes have now made it all but inevitable that economics will subsume politics.

Other interdependence analysts propose a different mix of politics and economics. In their landmark work on interdependence theory, Robert Keohane and Joseph Nye argue that economic changes do not negate politics, but create a new type of politics.[34] Their alternative conception of interdependence is a synthesis of international relations realism and liberal IPE. While accepting the realist notion that military force can dominate economic relations, Keohane and Nye add the qualification that military power may be irrelevant to issues other than direct threats to vital national interests. American nuclear power is virtually meaningless in regard to international monetary policy. Moreover, power, especially in an economically interdependent world, is complex. The ability to affect certain outcomes at acceptable cost will vary from issue-area to issue-area. A country can dominate monetary affairs but may be more sensitive to harmful retaliation in trade matters.

States also limit the actual use of power by establishing "international regimes"—tacit or explicit agreements on acceptable behavior within a given issue area. The General Agreement on Tariffs and Trade (GATT) constrains the use of economic power in international trade with threats of costly counteraction against undue protectionism. What interdependence does, for Keohane and Nye, is to create new sources of power; it produces networks of "mutual asymmetric dependence," which statesmen must consider when defining and defending state interests. Interdependence redefines politics.

Although Keohane and Nye stop well short of pronouncing the state irrelevant, their synthesis has a distinctively liberal hue.[35] In a sense, they save Keynesianism from itself by offering a justification for international economic management based upon a more sophisticated theory of politics. Furthermore, while the idea of "regime" raises some question as to liberalism's traditional unit of analysis, it is consistent with methodological individualism. As Keohane demonstrates elsewhere, regime creation and maintenance are in accord with the assumption of rational egoism.[36] The regime concept also returns to the question of how principles and norms can limit power, a primary concern of earlier liberal theorists. Keohane and Nye's approach to interdependence has its flaws; it is notably less successful in explaining North–South relations.[37] Their work, nonetheless, is an important revision of IPE liberalism.

Variants of Neo-Marxism

Liberalism is not the only IPE theory to undergo significant revision. Marxist IPE is rife with internal disgreements on a number of key issues. The causes and mechanisms of imperialism have been reconsidered. The nature of state and class power have been the subject of pointed contention. To cover adequately all of these issues is beyond the introductory scope of this chapter. Competent surveys of much of this literature are already available.[38] Thus, the focus here will be on one controversy in Marxist IPE: the impact of imperialism on less developed countries (LDCs). This topic is appropriate for closer inspection insofar as dependency theory and its offspring clearly illustrate conceptual revision within the Marxist approach.

Early Dependency Theory

After World War II, the traditional Marxist contention that imperialism leads to capitalist development of LDCs is seriously challenged. As the twentieth century unfolds, capital does expand out of the advanced industrial countries, but this does not necessarily lead to industrialization in less developed regions. Paul Baran is among the first to point out this problem.[39] He argues that the dialectical interaction of foreign capital and indigenous social forces does not always result in the creation of a typical capitalist mode of production in LDCs. Rather, surplus is expropriated from these areas and exported back to the advanced capitalist countries, yielding "underdevelopment" in Africa, Asia, and Latin America. Baran's insight paves the way for dependency theory.

Early dependency theorists generally assume a world-systemic perspective.[40] Theories of imperialism, especially Bukharin's, also employ a systemic level of analysis, but much of this debate concentrates on one component of the capitalist world-system: the advanced industrial countries. By consistently focusing on the nature of the larger system, dependency theorists analytically extend the internal dynamics of the capitalist mode of production, as originally outlined by Marx, to the world as a whole. Their description and explanation of the position of underdeveloped countries within the capitalist world-system can be seen as analogous to the proletariat's situation vis-à-vis the capitalist. Underdevelopment becomes a *necessary* condition for the maintenance of the capitalist world system *as a whole*. Economic development in underdeveloped countries is as impossible as eliminating the "immiseration" of the working class.

Early dependency theory also argues that underdevelopment is largely the result of forces external to the "satellite" countries. This point follows the assertion that a single mode of production, capitalism, dominates the world. From this vantage point, LDC class structures are an extension of the world-system of capital. The only chance of breaking out of the plight of underdevelopment, therefore, lies in cutting ties with the advanced industrial countries. This "delinking" from the centers of capitalism is yet another revision of earlier theories of imperialism, and starkly contrasts with Marx's expectation that capitalism would transform the world.

Dependency theory's revision of Marxist IPE, however, faces a new anomaly. Marxist Bill Warren puts forth the counterpoint that industrialization *has* occurred in some underdeveloped countries.[41] He suggests that these instances cannot be treated as "deviant cases" or exceptions that prove the rule of underdevelopment. Dependency theory, as such, is flawed, since it cannot explain outcomes contrary to its theoretical expectations.

Revisions of Dependency Theory

Two types of answers are given by Marxist IPE to meet the anomaly of actual LDC industrialization. On the one hand, some scholars return to Baran and look at the dialectical interaction of foreign capital and domestic social forces. Fernando Cardoso suggests that the particular "historical-structural" conditions that exist in a given LDC may give rise to capitalist development when foreign capital is introduced.[42] But this sort of development is not equated with the process that occurred earlier in advanced industrial countries. Instead, the genesis of capitalist production in the periphery is "associated dependent development." By this argument, industrialization in LDCs occurs only in historically specific instances of a mutually advantageous interrelationship between foreign capital, domestic capital, and the indigenous state. Other scholars move in a direction similar to Cardoso, arguing that a number of modes of production may exist simultaneously in the world system.[43] This has spawned a "modes of production" literature, which emphasizes domestic determinants of development and underdevelopment in the Third World. These studies move away from the international level of analysis and engage in more detailed comparative political economy research.

A second neo-Marxist answer to the "Warren anomaly" is the world-systems approach. This perspective reinforces the original level of analysis of dependency theory.[44] It reasserts the impossibility of numerous coincident modes of production; capitalism is the single world mode of production. World-systems analysis, however, modifies the dependency argument by confirming the existence of a "middle class" of states. Immanuel Wallerstein's "semiperiphery" attempts to capture the essence of this argument. The economic growth and political stablity of the system *as a whole* require a "middle sector" of countries. Thus, "limited opportunities for ascent" exist within the capitalist world-system. From this point of view, industrialization in LDCs is a natural outcome of the system.

Both of these revisions explain successful LDC industrialization in terms of Marxist IPE. They are similar insofar as both suggest only a limited number of LDCs will develop industrially. They are at odds over methodology. The historical-structuralist/modes of production perspective emphasizes the internal nature of the LDC in question, while world-systems analysis focuses on the character of the system as a whole. Both raise new questions; however, the anomaly is not completely solved.

The world-systems argument suffers from two weaknesses. First, the "semiperiphery" concept is indistinct. Because its existence is teleologically asserted, it is difficult to establish boundaries for the category.[45] A wide range of cases, with a host of apparent differences, are thrown together with little justification. Discussion, therefore, lapses into tautology, suggesting that a country is semiperipheral if it has some success with industrialization efforts. Secondly, the concept that may help overcome the vagueness of "semiperiphery," the state, is not adequately specified. The "strategies of ascent" for peripheral countries seeking development discussed by Wallerstein assume a measure of state autonomy and capacity. But these qualities are not defined. The world-systems argument rests on a simplistic assumption that "strong states" naturally exist in economically successful countries and "weak states" in the less fortunate. Critics have adduced counterexamples that challenge such statements.[46]

The domestic orientation of historical-structuralist/modes of production theory is explicitly aimed at explaining specific successes or failures in LDC industrialization. However, it too is unable to follow through. Cardoso, for example, emphasizes *political* dimensions of development but cannot escape the limitations of structuralist neo-Marxist theories of the state. Arguing that the state is a crucial element in the industrialization efforts of Latin American countries, he does not acknowlege the state's *potential* political autonomy. Instead, the state merely reflects the class structure of society.[47] Yet he also suggests class analysis does not allow for a complete explanation of particular cases of development or underdevelopment.[48] Thus, he recognizes the analytic constraints of Marxist IPE but does not move outside of them.[49]

Prominent class analysts are aware of the problems inherent in a neo-Marxist conception of the state. Fred Block argues that state power ought to be considered sui generis, not necessarily reducible to class interests. He flatly asserts that structuralist neo-Marxism is reductionist.[50] Ralph Miliband reaches a similar conclusion and suggests that the class/state relationship is best considered a "partnership" in which the state is not necessarily "junior partner."[51] These conclusions point to a fundamental problem for Marxist IPE. If it is to explain the anomaly of successful LDC industrialization, Marxist IPE must reconsider the possibilities of autonomous political action and the role of the state. In so proceeding, however, core assumptions of historical materialism may be eviscerated; the distinctive character of Marxism may be lost.

Neo-Realist IPE

While Marxist theory faces a theoretical conundrum, neorealist IPE is enjoying a renaissance after decades of decline.[52] Although mercantilist ideas are overshadowed after the turn of the century by liberal and Marxist approaches, economic nationalism is never wholly subsumed. At the practical level, protectionism thrives. In terms of theory, elements of mercantilist IPE are maintained, in different forms, by Max Weber and international relations realism.

Precursors

Max Weber posits an interactive relationship between politics and economics. He neither denies the efficiency of the market nor rejects the power of social classes. The state, though, is at the center of Weber's political economy.[53] A monopoly of the legitimate means of coercion within a given territory, if effectively realized, gives those who control the state obvious political advantage over rival social forces. The state can, therefore, under certain historical conditions, dominate both economic markets and social classes. The development of the modern state is dependent on the existence of a relatively advanced capitalist economy, but not merely in the Marxian sense of superstructure. The market cannot exist without the state; states grow out of markets. The two are intertwined in a complex and historically specific manner.[54] Weber argues that economic processes are not distinct from social and political processes. On the contrary, as the title of his magnum opus, *Economy and Society,* implies, these factors are interactive and interdependent.

Weber is similar to mercantilists and the German Historical School in two respects. First, he recognizes the potency of nationalism as a driving force of economic activity. Unlike some liberal interdependence theorists, he sees economic internationalization as heightening nationalist sentiments.[55] Second, he analytically places economic phenomena into a social and political context. Weber is, however, uncomfortable with historicism's lack of abstract theory. His methodology of ideal types is an epistemological compromise of theory and history. They are a heuristic and predictive device, suggesting general trends that may be redirected by particular historical conditions. His framework synthesizes nomothetic generalization and historical narrative. Although he concentrates on domestic political and economic forces, his conceptualizations are relevant to international relations as well.[56] For theoretical and sociological reasons, though, Weber's work did not immediately give rise to a further development and refinement of mercantilist IPE.

The contending liberal and Marxist variants of IPE were deeply entrenched at the turn of the century. Their wide acceptance weakened the insinuation of a new theoretical approach. Market and class defined central intellectual debates; the state was by and large an afterthought. Additionally, a growing division of labor both between and within social science disciplines shattered the common ground of mercantilist IPE. Not only was Weber's analysis ignored, but ironically came to be associated with a perspective quite unlike his own.[57]

International relations realism also maintains mercantilist assumptions. Drawing on the same classical political philosophers who inspired the mercantilists, realists views international affairs as a struggle for power among individually sovereign states in an anarchic world environment. Their analytic focus is the distribution of "power" around the globe. Although power is seen as a complex phenomenon (including, inter alia, economic resources), its ultimate expression is a state's military capability. This suggests that realists tend to a narrowly politicist vision of the relationship between politics and economics. They ignore Weber's more complex ideas of political–economic interaction. Unlike the earlier mercantilists, realists do not give due consideration to the economic role of the state; they downplay the "low" politics of economic affairs while emphasizing the "high" politics of diplomacy and the use of force.[58] Many IPE topics are thus peripheral concerns of international relations realism. Moreover, when realists discuss economic issues, they are vulnerable to criticism similar to that which undermined early mercantilists. Simply put, their theory of economics is inadequate.

The Critiques of Liberalism and Marxism

Although efforts were made to "bring the state back in" to IPE by a number of creative scholars, a neorealist approach was not fully developed until the 1970s.[59] At that time, liberal and Marxist programs are seriously weakened by changing international events and persistent theoretical anomalies. The undoing of the liberal world economy (the fall of Bretton Woods, the OPEC oil shocks) and the continuing developmental failures of LDCs moved liberal IPE toward interdependence. Some analysts, however, are less tolerant of core liberal assumptions and begin to develop a unique line of inquiry. Robert Gilpin argues against what he called the "transnational ideologists" and expands the narrow political focus of international relations realism to include economic issues.[60]

Contrary to the then popular liberal belief that the state was declining in significance as an international economic actor, Gilpin contends that the political and strategic interests of states lie at the heart of much of the world's economic activity. He reasserts the state as the key unit of analysis. He suggests that the politicism of earlier realism is untenable because economic forces may constrain political action, just as political power shapes economic outcomes. Gilpin suggests that in the interactive relationship between politics and economics, the former is especially important. In a manner reminiscent of Schmoller, he states: "politics determines the framework of economic activity and channels it in directions which tend to serve the political objectives of dominant political groups and organizations."[61] Other scholars followed Gilpin's lead and contributed to the development of a state-oriented analytic framework as an alternative to interdependence theory.[62]

Within Marxist IPE, similar changes are underway. In the 1970s, the intransigence of early dependency theory gives way to the "new" dependency theory and world-systems analysis. But these new arguments continue to accept, and be constrained by, the structuralist Marxist conception of politics. Theda Skocpol's seminal work is a thoroughgoing critique of neo-Marxist political economy.[63] She demonstrates how Marxist insights are undermined by assumptions of the "relative autonomy" of the state. Her argument is, at once, sympathetic and devastating to Marxist arguments. She shows how state actions do not always reflect the interests of dominant social classes nor the long-term interests of capital. Her thesis returns to Weber and is aimed directly at core neo-Marxist assumptions. The later work of Block and Miliband are, consciously or not, rejoinders to her analysis. Skocpol's work has its roots in class analysis but moved away from the premises of neo-Marxism towards the statist alternative.

Thus, neorealist IPE in the late 1970s is transformed in reaction to the shortcomings of both liberal and Marxist arguments. A theoretical tradition extending back to the mercantilists is rebuilt upon the revision of international relations realism and a rereading of Weber. It should be noted that neorealist IPE theorists are heirs to the *empirical* tradition of mercantilism, historicism, and realism, which emphasizes the importance of state power in economic affairs. They do not necessarily advocate protectionism or other normative prescriptions offered by earlier nationalists.

Two Contemporary State-Oriented Approaches

At the core of neorealist IPE lies the premise that the state's semipermanent bureaucratic apparatus critically shapes economic processes on both the national and international levels.[64] The state obtains this importance by its very nature. In its ideal typical form, the state is the only political–economic institution that claims external sovereignty and internal political predominance. In reality, the state's sovereignty and authority are limited; it is precisely the gap between ideal and actual that forms the research agenda of state power theories. This approach does not assume that the political interests of the state will always dominate economic forces. Rather, it suggests a complex and interactive relationship between politics and economics. Through the analytic lens of the state, it seeks to specify conditions under which one may determine the other.

Nor does the neorealist perspective assume all states are alike. It conceives of the state as a pattern of institutional power relationships that result from historically specific political struggles. Particular organizational structures and capacities vary from one state to another; concrete definitions of state interests are diverse. However, all states, whatever their form, have as their primary purpose control of a certain territory. They are created to dominate domestic social forces and resist external threats. This essential similarity provides common ground for various arguments that fall under the rubric of the neorealist research program.

The focus on the state has given rise to research on the international and intranational levels of analysis. This is to be expected insofar as the state mediates global and domestic political–economic forces. From the international-systemic point of view, the analysis concentrates on the state's external relations with other states. This avenue of inquiry revises the "structural realism" of Kenneth Waltz.[65] Gilpin's conception of the international system, for example, incorporates both political and economic forces.[66] Statist analyses further consider how the international political–economic system influences the behavior of individual states. Gilpin holds that structural change, an economic as well as political process, may spark large-scale international war. Somewhat less drastically, Stephen Krasner suggests that a state facing hegemonic decline may pursue more nationally oriented policies.[67] Elsewhere, Krasner contends that less developed countries are motivated, in part, by their subordinate international structural position, to agitate for changes in international regimes.[68]

Neorealism also includes studies that focus on the character of domestic political structures and how they influence international economic policy. Early work along these lines suggests that the degree of centralization of state power and social forces significantly shapes the type of policy pursued.[69] More recent arguments concentrate on two characteristics of state–society relations: state autonomy and capacity.[70] Autonomy implies that the state elite has a distinct power base that allows it to formulate policy independent of particularistic interests or powerful social forces. Capacity suggests that even if the state enjoys autonomy, it may not be effective in action. It must have the administrative apparatus required to implement its preferred programs. Using these concepts, neorealist writers attempt to show how the specific character of state autonomy and capacity explain economic performance in both advanced industrial countries and less developed countries.

This raises the question of whether a stricter distinction should be made between the study of comparative and international political economy. Any attempt to rigidly enforce such separation would, however, weaken neorealist IPE. The two lines of neorealist thought, though operating on different levels of analysis, address many of the same empirical topics and complement one another. Moreover, placing the two approaches under the same theoretical rubric suggests that a broader synthesis may be in the offing.

Contemporary state-oriented approaches succeed in rehabilitating core nationalist assumptions and overcoming problems that had plagued mercantilism. They also contribute stimulating new ideas. These range from describing the conditions for world war to explicating the patterns of development in the Third World.[71] These arguments directly address the key problems confounding both liberalism and Marxism: why apparently autonomous political action influences economic processes. In its intranational manifestation, however, neorealism faces at least one drawback. Its methodology rests on "historical specificity" (i.e., an aversion to detailed definitions of autonomy and capacity because the concrete manifestations of each vary according to historical context), and this limits its ability to create generalizations. In addition, neorealism generally may pay insufficient attention to nonstate actors and processes of IPE. Whether these are serious weaknesses or not will be determined by further research.

Crossovers and Combinations

In sum, the three major schools of IPE research have all experienced significant revision and development. Each has inspired lively debate on basic premises and specific applications. Much of this theoretical evolution is the result of arguments internal to each paradigm. The different approaches have, however, also influenced one another. Contemporary state-oriented theories are founded upon critiques of liberalism and Marxism. Interdependence theory is engaged in a dialogue with neorealism. Marxist ideas are integrated into non-Marxist analysis. Cross-fertilization is so pervasive now that it has bred IPE approaches that are not easily placed within any of the three major theories. Rational choice analysis, hegemonic stability theory, and regime analysis are examples of such theoretical crossovers.

Rational Choice Analysis

Sometimes called the "new political economy," rational choice analysis is grounded in liberalism.[72] It constructs abstract models from the core liberal premise of rational egoism. From this assumption, arguments are developed to explain why individuals make choices that lead to economically suboptimal outcomes. In essence, it employs the logic of economic analysis to explain why Adam Smith's vision is not realized. The parsimony of this approach allows it to be used across levels of analysis.[73] For instance, the persistence of protectionism, a continuing weakness of liberal theory in general, has been the focus of at least three different rational choice arguments. Building upon the early work of Mancur Olson, some rational choice analysts view the maintenance of free trade as an international public good.[74] It involves a "free-rider" problem in which each state tries to avoid the costs of maintaining free trade in the knowledge that it cannot be excluded from its benefits. This argument rests on the assumption that states are unitary rational actors that carefully calculate the costs and benefits of international cooperation.

A second rational choice explanation of protectionism holds that it is best understood as a result of bargaining among domestic interest groups.[75] From this perspective, rationality operates at the group or organizational level. Quite simply, there is a domestic "political market" for tariffs with a demand side (industry interest groups) and a supply side (government and civil service). Tariffs and nontariff barriers are a function of the balance of these two forces. Thirdly, some rational choice theorists invoke game theory.[76] This perspective returns to an international level of analysis, where states are taken as unitary rational actors. From this point of view, protectionism is seen as the outcome of a Prisoner's Dilemma game.[77] In an effort to avoid the "suckers payoff" of eschewing protectionism while others practice it, and in the hope of gaining the highest possible individual payoff by maintaining protectionism while others reject it, states are slow to reduce tariff and nontariff barriers.

Although there are notable differences among these three arguments, they all rest on a similar theoretical base. Each is a logical extension of the assumption of rational egoism under a specific set of circumstances. In each case, the political sphere is seen to bear distinct similarities to the working of a competitive market. All share the assumption that exogenous forces do not constrain rational actors from pursuing their own self-interests.

The theoretical strength of rational choice analysis is limited, though, by its abstractness. It extends the ceteris paribus assumption to such extremes that its predictions can be falsified by a host of "exogenous" variables. Thus, while some consider rational choice analysis a "theory" in and of itself, others view it as an analytic technique within a larger theoretical system. In the liberal sphere, Keohane uses a rational choice approach as one element of his further development of interdependence theory.[78] Indeed, rational choice analysis contributes directly to liberal IPE theory by explaining why political intervention thwarts the realization of liberal expectations. Other analysts, working outside of liberalism, have also invoked it. Gilpin employs rational choice analysis to explicate why states define their interests internationally.[79] The stated purpose of this work is to extend international relations realism to IPE theory.[80] Rational choice has also found its way into some Marxist analyses.[81] In this manner, what is essentially a liberal analytic tool is used to bolster contending theories. Rational choice has clearly been appropriated by nonliberal theories.

Hegemonic Stability Theory

While rational choice analysis grows out of a single theory and is extended to others, hegemonic stability theory arises almost simultaneously from all three major schools of thought. The common theme is the correlation of a relatively open international economic system and a concentration of global political–economic power in a single "hegemonic" state. The hegemon creates and defends a particular international economic order by absorbing costs of system maintenance and by compelling other states to accept certain practices. Under such circumstances, the worst excesses of debilitating economic nationalism are restrained by the hegemonic state. The United States and Great Britain have, at different historical moments, provided such leadership. Each has also faced gradual and relative economic decline, which theorists believe increases the likelihood of international economic conflict. All forms of hegemonic stability theory explain structural political requirements for orderly and efficient international economic relations.

Differences exist, however, regarding the meaning and significance of hegemonic power. Charles Kindleberger's liberal interpretation suggests that the free rider problem in global economic management requires a hegemonic power to provide public goods.[82] Without a hegemonic leader, economic nationalism could degenerate into international economic crisis as happened in the 1930s. This implies the hegemonic power must rise above the fray and serve the interest of the system as a whole. Other liberals, such as Robert Keohane, take another tack, arguing that hegemons act out of self-interest. The most powerful state creates specific international regimes to defend its interests systemwide. But he holds that regimes, and therefore a certain economic order, will persist after a hegemonic power has declined, because rationally egoistic states will perceive the maintenance of regimes to be in their interest.[83]

Immanuel Wallerstein offers a different view of hegemonic rise and decline. His historical analysis examines the origins of hegemonic power. Uneven development and long-term cycles of world growth cause the locus of accumulation to shift periodically from one region to another. A conjunction of such historical forces produces a hegemon at certain times. Once established, a hegemonic state's interests are intricately bound up with the preservation of the global system and it will act accordingly to protect these systemic interests. The hegemon thus enforces a particular division of labor among advanced industrial countries as well as between core and periphery. Moreover, since it must defend the capitalist system as a whole, it works to liberalize and open international economic relations generally. For Wallerstein, hegemonic power is less a matter of overcoming a "free rider" problem, or of providing specific regimes; it is an integral part of capitalism as a world-system.[84]

Neorealists also add to hegemonic stability theory. Krasner develops a "state-power" theory of hegemony in contradistinction to liberal interdependence IPE of the early 1970s.[85] He illustrates how states act upon interests defined in terms of economic power. Unlike Wallerstein's emphasis on the logic of capitalism, Krasner suggests the rise of a hegemonic power is the result of political struggles among competing states. On the other hand, Krasner agrees with Wallerstein that the leading state preceives its interests in systemic terms and therefore enforces a relatively liberal international economic order. When the hegemon's power fades, protectionism and economic nationalism revive. From another angle, where liberals emphasize the economic logic of hegemony, Krasner focuses on specific political interests. In an alternative neorealist approach, Gilpin employs rational choice analysis, grounded in liberalism, to explain a classically realist topic: the causes of war.[86] What unites Krasner and Gilpin, though, is the understanding that hegemonic power is a function of the competitive nature of interstate politics.

Thus, while the varieties of hegemonic stability theory are similar in their empirical observations, they differ in their interpretations and methodologies. Even among analysts working within one paradigm, neorealism for instance, there is disagreement on key questions. This diversity, moreover, has engendered criticism of hegemonic stability theory. Timothy McKeown notes that in the nineteenth century the United Kingdom did not act as the theory predicts.[87] Others question the extent to which the United States has actually declined in the past decades.[88] Keohane, who accepts the point of U.S. decline, shows how international regimes are unevenly affected by a hegemon's demise. He concludes that hegemonic stability theory alone does not explain regime change and turns scholarly attention toward a more detailed analysis of international regimes.

Regime Analysis

The study of international regimes is another offshoot of liberal IPE. It is a central feature of Keohane and Nye's interdependence theory. But much like rational choice, regime analysis has also been integrated into contemporary state-oriented IPE. The regime concept thus draws together IPE paradigms. While such convergence opens promising new avenues of research, it also creates definitional problems. In a survey of the literature, Stephen Haggard and Beth Simmons uncover three meanings of the term "regime," ranging from a very broad notion of patterned international behavior to a more restrictive idea of multilateral agreements.[89] Perhaps the most widely recognized definition is that formulated in a volume edited by Krasner. Here, international regimes are viewed as networks of "principles, norms, rules and decision-making procedures around which actor expectations converge in a given issue-area."[90] Yet even if many liberals and neorealists accept this definition, other controversies remain.

Debate rages over the significance of international regimes. How do such networks of principles and norms influence the behavior of states and other transnational actors? Some IPE theorists dismiss the concept out of hand, arguing that principles and norms are tenuous reflections of more basic power relationships.[91] A number of liberals take the opposite position and see regimes as basic elements in the world political economy; regimes are found virtually everywhere and they can influence state action.[92] Between these two extremes, other political economists accept that regimes have only a limited impact on world political economic relations. Some, though not all, of the work in this latter category falls within the liberal versus neorealist dichotomy.

On the liberal side, Keohane develops a "functionalist" theory of regimes based upon a notion of "political market failure" derived from neoclassical economics.[93] Although he accepts the proposition that regimes are usually created by self-interested hegemonic powers, Keohane's functionalism suggests regimes may come to possess a certain degree of autonomy vis-á-vis the interests of individual states. States abide by established regimes because violation in the context of interdependence entails national economic costs. Moreover, states seek to build regimes in the absence of a hegemonic power because their interests are best served by this form of international cooperation. Political interest is the initial impetus of economic cooperation, but the exigencies of interdependence ultimately constrain political interests. Alternatively, Stephen Krasner incorporates the regime concept into his neorealist theory of "structural conflict."[94] While recognizing a modicum of regime autonomy, his "modified realist" approach views regimes as reflections of the international distribution of power. Less developed countries try to change regimes because this is the most effective way for them to protect their national interests. For Krasner, regimes do not negate power politics; they redefine political strategies.

Krasner and Keohane, while they agree that regimes exist and can influence international relations, use contending theories to reach disparate conclusions on the significance of regimes. The broader debate on regimes, however, cannot be confined to differences between liberals and neorealists. Other neorealists, for example, either do not use the concept or reject it outright. In addition, liberals do not have a uniform approach to regime issues. Haggard and Simmons explicate four theories of international regimes, each with a distinct methodology. Some of these flow from liberal analysis (functionalism); others (game theory and structuralism) are consistent with liberalism but are employed by neorealists as well. Thus, while regime analysis needs further development, especially in determining precisely how regimes may affect state action, it has grown into a unique subfield of IPE.

Conclusion

To summarize, IPE theory has evolved along several fronts. Within the three major theoretical traditions, significant change has occurred. Liberalism has moved both backward and forward; neoclassical economics fails as an IPE theory because it ignores questions of power, but interdependence theory explicitly focuses on the interaction of economics and politics. Contemporary Marxism is characterized by lively debate and frequent revision. Marxist theory, however, now faces a serious problem in that necessary conceptual change may undermine core premises of historical materialism. State-oriented theory, on the other hand, has overcome the weaknesses of its mercantilist forebears and is enjoying a renaissance of sorts. IPE theorizing has also moved beyond the three central schools of thought. Recent work in rational choice analysis, hegemonic stability theory, and regime analysis offers new avenues of inquiry. In short, IPE theories are continually being reconsidered and reformulated.

The theoretical evolution of IPE theory has not been perfectly linear and cumulative. Progress is evident in that outmoded arguments have been discarded (e.g., the mercantilist definition of wealth as specie). However, new ideas have arisen from shifting historical circumstances as well as theoretical falsification. One strand of the multifaceted debate among Marxists has been inspired by the complex patterns of post-World War II development in the Third World. Revisions of liberalism and realism have been sparked by upheaval in the global political economy in the 1970s. Theoretical change has therefore come in discontinuous fits and starts. For decades, mercantilist-realist IPE stagnated, until it was revived in the 1970s. Similarly, Marxism and liberalism have experienced periods of uninspired rehearsals of well-worn generalities, punctuated by controversial and original revisions. Thomas Kuhn's notion of paradigmatic transformation, moving from "normal science" to "scientific revolution," seems to be borne out of the development of IPE theory.[95]

In looking ahead, it seems unlikely that any grand synthesis of the three major schools of thought is, or indeed can be, imminent. Ontological and epistemological differences continue to pose obstacles to more thoroughgoing theoretical melding. Moreover, the crossovers that have been attempted thus far are more syncretic than synthetic, involving uneasy combinations of potentially contradictory elements. Does the methodological individualism of rational choice analysis undermine the theoretical bases of Marxism and neorealism? Analysts have handled such questions in a pragmatic manner, implying that if greater explanatory precision and more interesting arguments emerge from their eclecticism, then theoretical consistency may well be the hobgoblin of little minds. Although serious epistemological issues remain, theoretical *bricolage* will likely endure.[96]

Notes

1. Robert Keohane, *After Hegemony* (Princeton: Princeton University Press, 1984), p. 21. Robert Gilpin goes further when he states that IPE questions: ". . . how the state and its associated political processes affect the production and distribution of wealth and, in particular, how political decisions and interests influence the location of economic activities and the distribution of the costs and benefits of these activities. Conversely, these questions also inquire about the effect of markets and economic forces on the distribution of power and welfare among states and other political actors, and particularly about how these economic forces alter the international distribution of political and military power." Robert Gilpin, *The Political Economy of International Relations* (Princeton: Princeton University Press, 1987), pp. 10–11.

2. Roger Tooze, "Perspectives and Theory: A Consumer's Guide," in Susan Strange, ed., *Paths to International Political Economy* (London: Allen and Unwin, 1984), p. 2.

3. Hans Morgenthau, a leading realist theoretician, was quite straightforward regarding the necessity of analytically separating politics and economics: "Intellectually, the political realist maintains the autonomy of the political sphere, as the economist, the lawyer, the moralist maintain theirs. . . . It is exactly through such a process of emancipation from other standards of thought, and the development of one appropriate to its subject matter, that economics has developed as an autonomous theory of the economic activity of man. To contribute to a similar development in the field of politics is indeed the purpose of political realism." Hans Morgenthau, *Politics Among Nations* (New York: Knopf, 1985) pp. 13–16.

4. Susan Strange, "International Political Economy: The Story So Far and the Way Ahead," in W. Ladd Hollist and F. Lamond Tullis, eds., *An International Political Economy* (Boulder, Col.: Westview, 1985).

5. Kenneth N. Waltz, *Theory of International Politics* (Reading, Mass.: Addison-Wesley, 1979).

6. Robert Gilpin, *The Political Economy of International Relations* (Princeton: Princeton University Press, 1987), chapter 2. An interesting counterpart to this view of the literature is Alfred Stepan's three categories of theory in *The State and Society: Peru in Comparative Perspective* (Princeton: Princeton University Press, 1978), chap. 1. The influence of Gilpin's analysis can be seen in Jeffery Frieden and David Lake, eds., *International Political Economy* (New York: St. Martin's Press, 1987). Stephen Gill and David Law also explicate the three major theories in *The Global Political Economy* (Baltimore: Johns Hopkins University Press, 1988). For other categorizations, see David H. Blake and Robert S. Walters, *The Politics of Global Economic Relations* (Englewood Cliffs, N.J.: Prentice-Hall, 1987); Roger Tooze, "Perspectives and Theory: A Consumer's Guide," in Susan Strange, ed., *Paths to International Political Economy* (London: Allen and Unwin, 1984); and R.J. Barry Jones, "Perspectives on International Political Economy," in Jones, ed., *Perspectives on Political Economy* (New York: St. Martin's Press, 1983).

7. Gilpin, 1987, op. cit., p. 26.

8. "If pressed I would describe myself as a liberal in a realist world and frequently even a world of Marxist class struggle." Robert Gilpin, "The Richness of the Tradition of Political Realism," *International Organization,* 38,2 (Spring 1984), p. 289.

9. Jacob Viner, "Power versus Plenty as Objectives of Foreign Policy in the Seventeenth and Eighteenth Centuries," *World Politics* 1,1 (October 1948).

10. Eli Heckscher, *Mercantilism* (N.Y.: Macmillan, 1955), p. 15.

11. N. Keohane, *Philosophy and the State in France* (Princeton: Princeton University Press, 1980). Karl Pribram offers an analysis of different forms of mercantilism throughout Europe in his *A History of Economic Reasoning* (Baltimore: Johns Hopkins University Press, 1983), pp. 31–89.

12. "In every phase of economic development, a guiding and controlling part belongs to some one or another political organ. . . . It rules economic life as well as political, determines its structure and institutions, and furnishes, as it were, the centre of gravity of the whole mass of socio-economic arrangements." Gustav Schmoller, *The Mercantile System and Its Historical Significance* (New York: Peter Smith, 1931 [1895]), p. 2.

13. Schumpeter's discussion of Schmoller is informative, and Staniland points to the politicist followers of the Historical School. Joseph Schumpeter, *History of Economic Analysis* (New York: Oxford University Press, 1954), pp. 809–815. See also Martin Staniland, *What Is Political Economy?* (New Haven: Yale University Press, 1985), pp. 83–86.

14. For the paradigmatic importance of Smith, see Phyllis Deane, *The Evolution of Economic Ideas* (Cambridge: Cambridge University Press, 1978), pp. 6–7.

15. Schumpeter captures this nicely: "The comprehensive vision of the universal interdependence of all the elements of the economic system . . . probably never cost Ricardo as much as an hour's sleep. His interest was in the clear-cut result of direct, practical significance. In order to get this he cut that general system to pieces, bundled up as large parts of it as possible, and put them in cold storage—so that as many things as possible should be frozen and given. He then piled one simplifying assumption upon another until, having really settled everything by these assumptions, he was left with only a few aggregate variables, between which, given these assumptions, he set up simple one-way relations so that, in the end, the desired results emerged almost as tautologies." Schumpeter, op. cit., pp. 472–73. Ricardo's methodological break with Smith is also discussed in Deane, op. cit., pp. 74–80.

16. "But if economics is essentially an engine of analysis, a method of thinking rather than a body of substantive results, Ricardo literally invented the technique of economics." Mark Blaug, *Economic Theory in Retrospect* (Cambridge: Cambridge University Press, 1985), pp 135–36.

17. Edmund Silberner, *The Problem of War in Nineteenth Century Economic Thought* (Princeton: Princeton University Press, 1946), p. 280; other quotations in this paragraph come from the same source, pp. 280–82.

18. Norman Angell, *The Great Illusion* (New York: G.P. Putnam, 1913).

19. Edward H. Carr, *The Twenty Years Crisis, 1919–1939* (London: Macmillan, 1951), pp. 43–46.

20. Many of these can be found in Shlomo Avineri, ed., *Karl Marx on Colonialism and Modernization* (Garden City, N.Y.: Doubleday, 1969).

21. Karl Marx, *Capital,* vol. I (New York: Vantage, 1976 [1867]), chap. 24.

22. "The bourgeoisie, by rapid improvement of all instruments of production, by the immensely facilitated means of communication, draws all, even the most barbarian, nations into civilisation. The cheap prices of its commodities are the heavy artil-

lery with which it batters down all Chinese walls, with which it forces the barbarians' intensely obstinate hatred of foreigners to capitulate. It compels all nations, on pain of extinction, to adopt the bourgeois mode of production; it compels them to introduce what it calls civilisation into their midst, i.e. to become bourgeois themselves. In one word, it creates a world after its own image." From the *Communist Manifesto,* as quoted in Avineri, op. cit. pp. 36–37.

23. Marx, op. cit., 1976, preface.

24. Theodor Shanin, ed., *Late Marx and the Russian Road* (New York: Monthly Review Press, 1983).

25. An excellent review of all of these theorists can be found in Anthony Brewer, *Marxist Theories of Imperialism* (London: Routledge & Kegan Paul, 1980).

26. Lenin, V.I. *Imperialism: The Highest Stage of Capitalism* (Beijing: Foreign Languages Press, 1975), p. 76.

27. Deane, op. cit., pp. 140–41, argues that Marx had little impact on mainstream economists.

28. Keynes states his position at the very beginning of the *General Theory:* "I shall argue that the postulates of the classical theory are applicable to a special case only and not to the general case, the situation which it assumes being a limiting point of the possible positions of equilibrium. Moreover, the characteristics of the special case assumed by the classical theory happen not to be those of the economic society in which we actually live, with the result that its teaching is misleading and disasterous if we attempt to apply it to the facts of experience." John M. Keynes, *The General Theory of Employment Interest and Money* (New York: Harcourt, Brace and Company, 1936), p. 3.

29. Mark Blaug, "Kuhn versus Lakatos, or paradigms versus research programs in the history of economics," *History of Political Economy* 7:4 (1975), p. 412.

30. This tendency can be seen in: John M. Keynes, "National Self-Sufficiency," *The Yale Review,* 22, 4 (June 1933): 755–69.

31. Milton Friedman, "The Case for Flexible Exchange Rates," in Milton Friedman, *Essays in Positive Economics* (Chicago: University of Chicago Press, 1953).

32. Susan Strange, op. cit., 1985.

33. This is, of course, a reference to Raymond Vernon's seminal work, *Sovereignty at Bay* (New York: Basic Books, 1971). Charles Kindleberger anticipated this analysis when he wrote: "the nation state is just about through as an economic unit." Charles Kindleberger, *American Business Abroad: Six Lectures on Direct Investment* (New Haven: Yale University Press, 1969), p. 209.

34. Robert Keohane and Joseph Nye, *Power and Interdependence: World Politics in Transition* (New York: Little, Brown, 1977).

35. ". . . although our analysis was clearly rooted in interdependence theory, which shared key assumptions of liberalism, we made no efforts to locate ourselves with respect to the liberal tradition. As we now see the matter, we were seeking in part to broaden the neofunctional strand of liberalism that had been developed by Ernst B. Haas and others in the 1950s and 1960s but that had been limited largely to the analysis of regional integration." Robert Keohane and Joseph Nye, *"Power and Interdependence* Revisited," *International Organization* 41,4 (Autumn 1987), p. 729.

36. Robert Keohane, *After Hegemony* (Princeton: Princeton University Press, 1984).

37. Michael Doyle, "Stalemate in the North–South Debate: Strategies and the New International Economic Order," *World Politics* 35,3 (April 1983); Vincent Mahler, "The political economy of North–South commodity bargaining: the case of

the International Sugar Agreement," *International Organization* 38,4 (Autumn 1984); John Ravenhill, "What is to be done for Third World commodity exporters? An evolution of the STABEX scheme," *International Organization* 38,3 (Summer 1984).

38. Brewer, op. cit.; Martin Carnoy, *The State and Political Theory* (Princeton, Princeton University Press, 1984; David A. Gold, Clarence Y.H. Lo, and Erik Olin Wright, "Recent Developments in Marxist Theories of the Capitalist State," *Monthly Review* 27 (1975), no. 5:29–43; no. 6:36–51.

39. Paul Baran, *The Political Economy of Growth* (New York: Monthly Review Press, 1957).

40. Andre Gunder Frank, *Capitalism and Underdevelopment in Latin America* (New York: Monthly Review Press, 1967); Theotonio Dos Santos, "The Structure of Dependence," *The American Economic Review* 60,2 (May 1970).

41. Bill Warren, "Imperialism and Capitalist Industrialization," *New Left Review,* 81 (September–October, 1973).

42. Fernando H. Cardoso, "Associated-Dependent Development: Theoretical and Political Implications," in Alfred Stepan, ed., *Authoritarian Brazil* (New Haven: Yale University Press, 1973), pp. 142–78; Fernando H. Cardoso and Enzo Faletto, *Dependency and Development in Latin America* (Berkeley: University of California Press, 1979). It is important to note that Cardoso's work develops simultaneous to Frank's dependency theory. Thus, he "solves" the anomaly of LDC industrialization even as the question is taking shape; see Kate Manzo, "Modernist Discourse and the Crisis of Development Theory," unpublished manuscript, 1989.

43. Ronald H. Chilcote and Dale L. Johnson, eds., *Theories of Development: Mode of Production or Dependency?* (Beverly Hill: Sage Publications, 1983).

44. Immanuel Wallerstein, *The Capitalist World-Economy* (New York: Cambridge University Press, 1979).

45. Theda Skocpol, "Wallerstein's World Capitalist System: A Theoretical and Historical Critique," *American Journal of Sociology* 82,5 (March 1977), pp. 1075–90.

46. *Ibid.*

47. "If the state has expanded and fortified itself, it has done so as the expression of a class situation which has incorporated capitalist development, as we have said, and policies of the dominant classes favorable to the rapid growth of the corporate system, to alliances between the state and business enterprises, and to the establishment of interconnections, at the level of the state productive system, between public and multinational enterprises." Cardoso and Faletto, op. cit., p. 201.

48. "A simple 'structural' analysis, demonstrating the contradictions between social forces and the drawbacks of the process of accumulation with its cycles and crises is insufficient to explain the concrete course of political events." *Ibid.*, p. 207.

49. The early work of Peter Evans is a broader synthesis of Cardoso's perspective and world-systems analysis in an effort to rejuvinate Marxists IPE. Peter Evans, *Dependent Development: The Alliance of Multinational, State, and Local Capital in Brazil* (Princeton: Princeton University Press, 1979).

50. In a provocative riposte to the structuralist definition of the state as the "condensation of class interests," Block argues that: "A condensation cannot exercise power." Fred Block, "Beyond Relative Autonomy: State Managers as Historical Subjects," *The Socialist Register, 1980* (London: Merlin Press, 1980), pp. 227–42.

51. Ralph Miliband, "State Power and Class Interests," *New Left Review* (March–April 1983).

52. The term "neorealist" is used here to avoid the normative implications of "nationalism."

53. Bertrand Badie and Pierre Birnbaum, *The Sociology of the State* (Chicago: University of Chicago Press, 1983), p. 17.

54. "On the one hand, capitalism in its modern stages of development requires the bureaucracy, though both have arisen from different historical sources. Conversely, capitalism is the most rational economic basis for bureaucratic administration and enables it to develop in the most rational form, especially because, from a fiscal point of view, it supplies the necessary money resources." Max Weber, *Economy and Society,* Guenther Roth and Claus Wittich, eds. (Berkeley: University of California Press, 1978), p. 224.

55. David Betham, *Max Weber and the Theory of Modern Politics* (Cambridge: Polity Press, 1985), p. 131.

56. Randall Collins offers a careful explication of Weber's significance for international relations; see Randall Collins, *Weberian Sociological Theory* (Cambridge: Cambridge University Press, 1986), pp. 145–66.

57. Weber's work was appropriated by functionalist sociologists to bolster a decidedly society-centered theoretical perspective, a use since criticized as untrue to Weber's own inclinations; see Badie and Birnbaum, op. cit., chap. 2.

58. Gilpin argues that only after World War II did realism lose sight of economic factors. Classical precursors of realist thought had a highly developed notion of political economy. Gilpin, 1984, op. cit., pp. 293–97.

59. Four authors stand out in this regard: Karl Polanyi, *The Great Transformation* (New York, 1944); Alexander Gershenkron, *Economic Backwardness in Historical Perspective* (Cambridge: Belknap Press, 1966); Gunnar Myrdal, *Asian Drama (New York: Vintage, 1970);* J.P. Nettl, "The State as a Conceptual Variable," *World Politics* 20,4 (July 1968), pp. 559–92.

60. Robert Gilpin, "The Politics of Transnational Economic Relations," in Robert Keohane and Joseph Nye, eds., *Transnational Relations and World Politics* (Cambridge: Harvard University Press, 1972).

61. *Ibid.,* p. 403.

62. Perhaps the most significant work along these lines is that of Stephen Krasner: Stephen D. Krasner, *Defending the National Interest* (Princeton: Princeton University Press, 1978); and "State Power and the Structure of International Trade," *World Politics* 28 (April 1976): 317–47.

63. Theda Skocpl, *States and Social Revolutions* (New York: Cambridge University Press, 1979).

64. Theda Skocpol, "Bringing the State Back In: Strategies of Analysis in Current Research," in Evans, et al., *Bringing The State Back In* (New York: Cambridge University Press, 1985).

65. Kenneth Waltz, op. cit.

66. Robert Gilpin, *War and Change in World Politics* (Princeton; Princeton University Press, 1981).

67. Stephen Krasner, "American Policy and Global Economic Stability," in Avery and Rapkin, eds. *America in a Changing World Economy* (New York: Longman, 1982).

68. Stephen Krasner, *Structural Conflict* (Berkeley: University of California Press, 1985).

69. Peter Katzenstein, ed., *Between Power and Plenty* (Madison: University of Wisconsin Press, 1978).

70. Peter Evans, et al., op. cit.

71. Gilpin discusses conditions leading to war in *War and Change in World Politics,* op. cit. Atul Kohli presents a number of analyses of Third World development in *The*

State and Development in the Third World (Princeton: Princeton University Press, 1986).

72. For a general discussion of "new political economy," see Martin Staniland, *What Is Political Economy,* op. cit., chap. 3. Although there are differences in their analytic focus and policy prescriptions, both game theory and public goods theory are here considered two variants of rational choice theory in that they operate on a similar level of abstraction and share many of the same assumptions. For distinctions between game theory and public goods theory, see John A.C. Conybeare, "Public Goods, Prisoner's Dilemmas and the International Political Economy," *International Studies Quarterly* 28,1 (March 1984), pp. 5–22.

73. Bruno Frey offers an explication and defense of rational choice approaches to international political economy in *International Political Economics* (New York: Basil Blackwell, 1984); and "The Public Choice View of International Political Economy," *International Organization,* 38,1 (Winter 1984). The inspiration for much of the rational choice theory in IPE is Mancur Olson. His most recent work is possibly the most far-reaching application of this approach to IPE topics; see Mancur Olson, *The Rise and Decline of Nations* (New Haven: Yale University Press, 1982).

74. Charles Kindleberger, *The World in Depression* (Berkeley: University of California Press, 1973).

75. Bruno Frey, *International Political Economics,* op. cit.

76. John A.C. Conybeare, "Public Goods, Prisoner's Dilemmas and the International Political Economy," op. cit.

77. Game theory has also been applied to topics of international political economy in Kenneth Oye, ed., *Cooperation Under Anarchy* (Princeton: Princeton University Press, 1986).

78. Keohane, *After Hegemony,* op. cit.

79. Robert Gilpin, *War and Change in World Politics,* op. cit.

80. Robert Gilpin, *The Political Economy of International Relations,* op. cit., p. xiii.

81. Good introductions to this literature are: John Roemer, " 'Rational Choice' Marxism: Some Issues of Method and Substance," and Jon Elster, "Further Thoughts on Marxism, Functionalism, and Game Theory," both in John Roemer, ed., *Analytical Marxism* (Cambridge: Cambridge University Press, 1986). For a critique of "analytical Marxism" from a dialectical-Marxist perspective, see Andrew Levine, et al., "Marxism and Methodological Individualism," *New Left Review,* no. 162 (March–April 1987).

82. Charles Kindleberger, one of the earliest contributors to hegemonic stability theory, argues that: "for the world economy to be stabilized, there has to be a stabilizer, one stabilizer." Kindleberger, 1973, op. cit., p. 305.

83. Robert Keohane, "The Theory of Hegemonic Stability and Changes in International Regimes, 1967–1977," in O. Holsti, et al., eds., *Change in the International System* (Boulder, Col: Westview, 1980).

84. A Gramscian perspective on "hegemony" can be found in Robert Cox, "Social Forces, States, and World Orders," *Millennium* (1981), vol. 10, pp. 126–55; and Gill and Law, op. cit., pp. 76–80.

85. Krasner, "State Power and the Structure of International Trade," op. cit.

86. Gilpin, *War and Change in World Politics,* op. cit.

87. Timothy J. McKeown, "Hegemonic Stability Theory and 19th Century Tariff Levels in Europe," *International Organization* 37:73–91.

88. Bruce Russett, "The Mysterious Case of Vanishing Hegemony: Or, Is Mark Twain Really Dead?" *International Organization* 39,2 (1985): 207–232; Susan Strange,

"The Persistant Myth of Lost Hegemony," *International Organization* 41,4 (1987): 551–74.

89. Stephan Haggard and Beth A. Simmons, "Theories of International Regimes," *International Organization* 41,3 (Summer 1987), pp. 491–517.

90. Stephen Krasner, ed., *International Regimes* (Ithaca: Cornell University Press, 1983).

91. Susan Strange's article in Krasner, 1983, op. cit., is the most pointed of these critiques.

92. Donald J. Puchala and Raymond F. Hopkins, "International Regimes: Lessons from Inductive Analysis," in Krasner, 1983, op. cit.

93. Robert Keohane, *After Hegemony,* op. cit.

94. Krasner, *Structural Conflict,* op. cit.

95. Thomas Kuhn, *The Structure of Scientific Revolutions* (Chicago: University of Chicago Press, 1962).

96. Claude Levi-Strauss uses the term *bricolage* to describe the cobbling together of diverse ideas in the process of myth-making. Recent crossovers in IPE theory are similarly composed of a hodgepodge of concepts and methods. See Claude Levi-Strauss, *The Savage Mind* (Chicago: University of Chicago Press, 1962), pp. 16–33).

Classical Mercantilism

Early Mercantilist political economy boldly asserts the centrality of politics. Economics is not separated from its political context, but instead is considered an important means of enhancing state power. Although the various strands of classical mercantilism do not form a unified theory, a general sense of the primacy of politics was the dominant view of political economy before Adam Smith, especially in the sixteenth and seventeenth centuries. The selections here, from Alexander Hamilton and Fredrich List, are not representative of the golden age of mercantilism; however, they both defend core mercantilist principles against the emerging liberal critique.

Alexander Hamilton (1755–1804) is most concerned with national economic development. His "Report on Manufactures," submitted to Congress in 1791, and his other analyses of public credit and a central bank are aimed at strengthening the flegdling economy of the United States. More of a practical statesman than social theorist, Hamilton invokes a number of classical mercantilist arguments in advocating the growth of manufacturing. Although he recognizes the importance of agriculture, he sees industry as providing additional dynamism to the American economy. Such diversification, in turn, will lessen the nation's vulnerability to external economic forces. The domestic market is a more reliable foundation upon which to build. Moreover, national security is best served by a self-reliant and complex national economy: "The extreme embarrassments of the United States in the late War, from an incapacity of supplying themselves, are still a matter of keen recollection . . ." Thus, avoiding economic dependence is an underlying theme, an idea consistent with earlier mercantilist thought.

The state, Hamilton argues, must take an active role in developing a manufacturing economy. Domestic obstacles to industrialization, such as inexperience and capital shortages, can only be overcome by government intervention: ". . . the public purse must supply the deficiency of private resource." Additionally, the state must promote and protect emerging U.S. industries, because this is common practice in the competitive international system. Adam Smith's notion of free trade is simply an inaccurate description of world political–economic realities. Hamilton, therefore, offers a series of protectionist measures that further American manufacturing interests. Yet his desire for protectionism differs from earlier mercantilist thought in that a positive balance of trade and an inflow of specie are not ends in themselves; they are the outcome of a strong manufacturing economy.

Hamilton thus draws on key mercantilist concepts (i.e., national security and self-reliance) but modifies others (the significance of a trade surplus). Interestingly, his rationale for U.S. economic policy is strikingly similar to the nationalist-oriented concerns of less developed countries in the twentieth century.

Fredrich List (1789–1846) operates more on the theoretical level. He directly assaults the liberal arguments of Smith and others of the "popular school." His critique centers on the invalidity of key liberal assumptions. The "cosmopolitical" world view assumes political stability and peace, even unity. This, List argues, misses the key point that the world is riven by pointed nationalist rivalries. Political economy must begin with a recognition of the inherently conflictual nature of international relations. The nation is, as early mercantilists held, the key unit of political–economic analysis. List's nationalist perspective reveals the politics behind international economic processes. "True political science," he contends, recognizes that the interests of the industrially developed economies are best served by a free trade regime; nations that have yet to industrialize are developmentally constrained in an open, competitive world economy. He concludes that protectionism is the necessary course for those countries interested in economic development.

Although List seems to offer a general theory of political economy, taking on Malthus as well as Smith, his argument is ultimately ad hoc. He suggests that "every separate nation" will discover its peculiar national means to economic development, but he is particularly concerned with Germany's economic future. Elsewhere in his book, *The National System of Political Economy,* he argues that the world can be divided into two regions, the "hot zone" and the "temperate zone." Only the latter is suitable for industrial development similar to that of Britain; the former should remain primarily agricultural. Thus, infant industry protection is not appropriate for every state, but only those deserving nations in the "temperate zone," such as Germany and the United States. Once these sorts of countries develop to a sufficient degree, then the mutual benefits of free trade can be realized. In sum, List recognizes the power of liberal arguments on comparative advantage and free trade. But he wants to make sure that, as these dynamics grow, Germany and others share in the wealth.

Report on Manufactures

ALEXANDER HAMILTON

The expediency of encouraging manufactures in the United States, which was not long since deemed very questionable, appears at this time to be pretty generally admitted. The embarrassments, which have obstructed the progress of our external trade, have led to serious reflections on the necessity of enlarging the sphere of our domestic commerce: the restrictive regulations, which in foreign markets abridge the vent of the increasing surplus of our Agricultural produce, serve to beget an earnest desire, that a more extensive demand for that surplus may be created at home: And the complete success, which has rewarded manufacturing enterprise, in some valuable branches, conspiring with the promising symptoms, which attend some less mature essays, in others, justify a hope, that the obstacles to the growth of this species of industry are less formidable than they were apprehended to be, and that it is not difficult to find, in its further extension, a full indemnification for any external disadvantages, which are or may be experienced, as well as an accession of resources, favorable to national independence and safety.

There still are, nevertheless, respectable patrons of opinions, unfriendly to the encouragement of manufactures. The following are, substantially, the arguments by, which these opinions are defended.

"In every country (say those who entertain them) Agriculture is the most beneficial and *productive* object of human industry. This position, generally, if not universally true, applies with peculiar emphasis to the United

This is the text that has been printed in Jacob E. Cooke (ed.) *The Reports of Alexander Hamilton* (N.Y.: Harper & Row, 1964).

States, on account of their immense tracts of fertile territory, uninhabited and unimproved. . . .

"To endeavor, by the extraordinary patronage of Government, to accelerate the growth of manufactures, is, in fact, to endeavor, by force and art, to transfer the natural current of industry from a more, to a less beneficial channel. Whatever has such a tendency must necessarily be unwise. Indeed it can hardly ever be wise in a government, to attempt to give a direction to the industry of its citizens. This under the quick-sighted guidance of private interest, will, if left to itself, infallibly find its own way to the most profitable employment: and it is by such employment, that the public prosperity will be most effectually promoted. To leave industry to itself, therefore, is, in almost every case, the soundest as well as the simplest policy.

"If contrary to the natural course of things, an unseasonable and premature spring can be given to certain fabrics, by heavy duties, prohibitions, bounties, or by other forced expedients; this will only be to sacrifice the interests of the community to those of particular classes. Besides the misdirection of labor, a virtual monopoly will be given to the persons employed on such fabrics; and an enhancement of price, the inevitable consequence of every monopoly, must be defrayed at the expense of the other parts of the society. It is far preferable, that these persons should be engaged in the cultivation of the earth, and that we should procure, in exchange for its productions, the commodities, with which foreigners are able to supply us in greater perfection, and upon better terms."

It ought readily to be conceded that the cultivation of the earth—as the primary and most certain source of national supply . . .—*has intrinsically a strong claim to pre-eminence over every other kind of industry.*

But, that it has a title to any thing like an exclusive predilection, in any country, ought to be admitted with great caution. That it is even more productive than every other branch of industry requires more evidence than has yet been given in support of the position. That its real interests, precious and important as without the help of exaggeration, they truly are, will be advanced, rather than injured by the due encouragement of manufactures, may, it is believed, be satisfactorily demonstrated. And it is also believed that the expediency of such encouragement in a general view may be shown to be recommended by the most cogent and persuasive motives of national policy.

The foregoing suggestions are *not designed to inculcate an opinion that manufacturing industry is more productive than that of Agriculture.* They are intended rather to show that the reverse of this proposition is not ascertained; that the general arguments which are brought to establish it are not satisfactory; and, consequently that a supposition of the superior productiveness of Tillage ought to be no obstacle to listening to any substantial inducements to the encouragement of manufactures, which may be otherwise perceived to exist, through an apprehension; that they may have a tendency to divert labour from a more to a less profitable employment.

. . . . But without contending for the superior productiveness of Manufacturing Industry, it may conduce to a better judgment of the policy, which ought to be pursued respecting its encouragement, to contemplate the subject, under some additional aspects, tending not only to confirm the idea, that this kind of industry has been improperly represented as unproductive in itself; but to evince in addition that the establishment and diffusion of manufactures have the effect of rendering the total mass of useful and productive labor, in a community, *greater than it would otherwise be*. . . .

. . . . [M]anufacturing establishments not only occasion a positive augmentation of the Produce and Revenue of the Society, but . . . they contribute essentially to rendering them greater than they could possibly be, without such establishments. These circumstances are. . . .

1). . . . The Division Of Labour

It has justly been observed, that there is scarcely any thing of greater moment in the economy of a nation than the proper division of labour. The separation of occupations causes each to be carried to a much greater perfection, than it could possibly acquire, if they were blended. . . .

. . . . [T]he mere separation of the occupation of the cultivator, from that of the Artificer, has the effect of augmenting the *productive powers* of labour, and with them, the total mass of the produce or revenue of a Country. In this single view of the subject, therefore, the utility of Artificers of Manufacturers, towards promoting an increase of productive industry, is apparent.

2). . . . An Extension Of The Use Of Machinery, A Point Which, Though Partly Anticipated Requires To Be Placed In One Or Two Additional Lights

The employment of Machinery forms an item of great importance in the general mass of national industry. . . . It shall be taken for granted, and the truth of the position referred to observation, that manufacturing pursuits are susceptible in a greater degree of the application of machinery, than those of Agriculture. If so all the difference is lost to a community, which, instead of manufacturing for itself, procures the fabrics requisite to its supply from other Countries. The substitution of foreign for domestic manufactures is a transfer to foreign nations of the advantages accruing from the employment of Machinery, in the modes in which it is capable of being employed, with most utility and to the greatest extent.

3). . . . The Additional Employment Of Classes Of The Community, Not Originally Engaged In The Particular Business

This is not among the least valuable of the means, by which manufacturing institutions contribute to augment the general stock of industry and production. In places where those institutions prevail, besides the persons regularly engaged in them, they afford occasional and extra employment to industrious individuals and families, who are willing to devote the leisure resulting from the intermissions of their ordinary pursuits to collateral labours, as a resource for multiplying their acquisitions or their enjoyments. . . .

. . . . It is worthy of particular remark, that, in general, women and children are rendered more useful, and the latter more early useful by manufacturing establishments than they would otherwise be. . . .

4). . . . Promoting Of Emigration From Foreign Countries

If it be true then, that it is the interest of the United States to open every possible avenue to emigration from abroad, it affords a weighty argument for the encouragement of manufactures; which, . . . will have the strongest tendency to multiply the inducements to it.

Here is perceived an important resource, not only for extending the population, and with it the useful and productive labour of the country, but likewise for the prosecution of manufactures, without deducting from the number of hands, which might otherwise be drawn to Tillage and even for the indemnification of Agriculture for such as might happen to be diverted from it. . . .

5). . . . Furnishing Greater Scope For The Diversity Of Talents And Dispositions, Which Discriminate Men From Each Other

This is a much more powerful means of augmenting the fund of national Industry than may at first sight appear. It is a just observation, that minds of the strongest and most active powers for their proper objects fall below mediocrity and labour without effect, if confined to uncongenial pursuits. And it is thence to be inferred, that the results of human exertion may be immensely increased by diversifying its objects. When all the different kinds of industry obtain in a community, each individual can find his proper element, and can call into activity the whole vigour of his nature. And the community is benefitted by the services of its respective members, in the manner, in which each can serve it with most effect.

6). . . . Affording A More Ample And Various Field For Enterprise

. . . . To cherish and stimulate the activity of the human mind, by multiplying the objects of enterprise, is not among the least considerable of the expedients, by which the wealth of a nation may be promoted. . . . The spirit of enterprise, useful and prolific as it is, must necessarily be contracted or expanded in proportion to the simplicity or variety of the occupations and productions, which are to be found in a Society. It must be less in a nation of mere cultivators, than in a nation of cultivators and merchants; less in a nation of cultivators and merchants, than in a nation of cultivators, artificers and merchants.

7). . . . Creating, In Some Instances, A New, And Securing In All A More Certain And Steady Demand, For Surplus Produce Of This Soil

It is evident, that the exertions of the husbandman will be steady or fluctuating, vigorous or feeble, in proportion to the steadiness or fluctuation, adequateness or inadequateness, of the markets on which he must depend, for the vent of the surplus, which may be produced by his labor; and that such surplus in the ordinary course of things will be greater or less in the same proportion.

For the purpose of this vent, a domestic market is greatly to be preferred to a foreign one; because it is in the nature of things, far more to be relied upon.

To secure such a market, there is no other expedient, than to promote manufacturing establishments. Manufacturers who constitute the most numerous class, after the Cultivators of land, are for that reason the principal consumers of the surplus of their labour.

The foregoing considerations seem sufficient to establish, as general propositions, that it is the interest of nations to diversify the industrious pursuits of the individuals who compose them—that the establishment of manufactures is calculated not only to increase the general stock of useful and productive labour; but even to improve the state of Agriculture in particular, certainly to advance the interest of those who are engaged in it. . . .

If the system of perfect liberty to industry and commerce were the prevailing system of nations, the arguments which dissuade a country, in the predicament of the United States, from the zealous pursuit of manufactures, would doubtless have great force. It will not be affirmed, that they might not be permitted, with few exceptions, to serve as a rule of national conduct. In such a state of things, each country would have the full benefit of its peculiar advantages to compensate for its deficiencies or disadvantages. If one nation were in a condition to supply manufactured articles on better terms than another, that other might find an abundant indemnification in a superior capacity to furnish the produce of the soil. And a free exchange, mutually beneficial, of the commodities, which each was able to supply, on the best terms, might be carried on between them, supporting in full vigour the industry of each. . . .

But the system which has been mentioned, is far from characterising the general policy of Nations. The prevalent one has been regulated by an opposite spirit. The consequence of it is, that the United States are to a certain extent in the situation of a country precluded from foreign Commerce. They can indeed, without difficulty obtain from abroad the manufactured supplies, of which they are in want; but they experience numerous and very injurious impediments to the emission and vent of their own commodities. Nor is this the case in reference to a single foreign nation only. The regulations of several countries, with which we have the most extensive intercourse, throw serious obstructions in the way of the principal staples of the United States.

In such a position of things, the United States cannot exchange with Europe on equal terms; and the want of reciprocity would render them the victim of a system which should induce them to confine their views to Agriculture, and refrain from Manufactures. A constant and increasing necessity, on their part, for the commodities of Europe, and only a partial and occasional demand for their own, in return, could not but expose them to a state of impoverishment, compared with the opulence to which their political and natural advantages authorize them to aspire.

 . . . . It is for the United States to consider by what means they can render themselves least dependent on the combinations, right or wrong of foreign policy.

The remaining objections to a particular encouragement of manufacturers in the United States now require to be examined.

One of these turns on the proposition, that Industry, if left to itself, will naturally find its way to the most useful and profitable employment: whence it is inferred that manufactures without the aid of government will grow up as soon and as fast, as the natural state of things and the interest of the community may require.

Against the solidity of this hypothesis, in the full latitude of the terms, very cogent reasons may be offered. These have relations to—the strong influence of habit and the spirit of imitation—the fear of want of success in untried enterprises—the intrinsic difficulties incident to first essays towards a competition with those who have previously attained to perfection in the business to be attempted—the bounties premiums and other artificial encouragements, with which foreign nations second the exertions of their own Citizens in the branches, in which they are to be rivalled.

Experience teaches, that men are often so much governed by what they are accustomed to see and practice, that the simplest and most obvious improvements, in the most ordinary occupations, are adopted with hesitation, reluctance, and by slow gradations. The spontaneous transition to new pursuits, in a community long habituated to different ones, may be expected to be attended with proportionably greater difficulty. When former occupations ceased to yield a profit adequate to the subsistence of their followers, or when there was an absolute deficiency of employment in them, owing to the superabundance of hands, changes would ensue; but these changes would be likely to be more tardy than might consist with the interest either of individuals or of the Society. In many cases they would not happen, while a bare support could be insured by an adherence to ancient courses; though a resort to a more profitable employment might be practicable. To produce the desireable changes as early as may be expedient, may therefore require the incitement and patronage of government.

The superiority antecedently enjoyed by nations, who have preoccupied and perfected a branch of industry, constitutes a more formidable obstacle, than either of those, which have been mentioned, to the introduction of the same branch into a country in which it did not before exist. To maintain between the recent establishments of one country and the long matured establishments of another country, a competition upon equal terms, both as to quality and price, is in most cases impracticable. The disparity, in the one or in the other, or in both, must necessarily be so considerable as to forbid a successful rivalship, without the extraordinary aid and protection of government.

But the greatest obstacle of all to the successful prosecution of a new branch of industry in a country, in which it was before unknown, consists, as far as the instances apply, in the bounties premiums and other aids which are granted, in a variety of cases, by the nations, in which the establishments to be imitated are previously introduced. It is well known (and particular examples in the course of this report will be cited) that certain nations grant bounties on the exportation of particular commodities, to enable their own workmen to undersell and supplant all competitors in the countries to which those commodities are sent. Hence the undertakers of a new manufacture have to contend not only with the natural disadvantages of a new undertaking, but with the gratuities and remunerations which other governments bestow. To be enabled to contend with success, it is evident that the interference and aid of their own government are indispensable.

There remains to be noticed an objection to the encouragement of manufactures, of a nature different from those which question the probability of success. This is derived from its supposed tendency to give a monopoly of advantages to particular classes, at the expense of the rest of the community, who, it is affirmed, would be able to procure the requisite supplies of manufactured articles on better terms from foreigners, than from our own Citizens, and who, it is alleged, are reduced to the necessity of paying an enhanced price for whatever they want, by every measure, which obstructs the free competition of foreign commodities.

But though it were true, that the immediate and certain effect of regulations controling the competition of foreign with domestic fabrics was an increase of Price, it is universally true, that the contrary is the ultimate effect with every successful manufacture. When a domestic manufacture has attained to perfection, and has engaged in the prosecution of it a competent number of Persons, it invariably becomes cheaper. Being free from the heavy charges which attend the importation of foreign commodities, it can be afforded, and accordingly seldom or never fails to be sold Cheaper, in process of time, than was the foreign Article for which it is a substitute. The internal competition which takes place, soon does away [with] every thing like Monopoly, and by degrees reduces the price of the Article to the *minimum* of a reasonable profit on the Capital employed. This accords with the reason of the thing, and with experience.

Whence it follows, that it is the interest of a community, with a view to eventual and permanent economy, to encourage the growth of manufactures. In a national view, a temporary enhancement of price must always be well compensated by permanent reduction of it.

There seems to be a moral certainty, that the trade of a country which is both manufacturing and Agricultural will be more lucrative and prosperous than that of a Country, which is merely Agricultural.

One reason for this is found in that general effort of nations . . . to procure from their own soils, the articles of prime necessity requisite to their own consumption and use, and which serves to render their demand for a foreign supply of such articles, in a great degree occasional and contingent. . . .

Another circumstance which gives a superiority of commercial advantages to states that manufacture as well as cultivate, consists in the more numerous attractions, which a more diversified market offers to foreign Customers, and in the greater scope which it affords to mercantile enterprise. . . .

From these circumstances collectively—two important inferences are to be drawn, one, that there is always a higher probability of a favorable balance of Trade, in regard to countries in which manufactures founded on the basis of a thriving Agriculture flourish, than in regard to those, which are confined wholly or almost wholly to Agriculture; the other (which is also a consequence of the first), that countries of the former description are likely to possess more pecuniary wealth, or money, than those of the latter.

 . . . . [T]he uniform appearance of an abundance of specie, as the concomitant of a flourishing state of manufactures, and of the reverse, where they do not prevail, afford a strong presumption of their favorable operation upon the wealth of a Country.

Not only the wealth, but the independence and security of a country, appear to be materially connected with the prosperity of manufactures. Every nation, with a view to those great objects, ought to endeavour to possess within itself all the essentials of national supply. These comprise the means of *Subsistence, habitation, clothing,* and *defence.*

The possession of these is necessary to the perfection of the body politic; to the safety as well as to the welfare of the society; the want of either is the want of an important Organ of political life and Motion; and in the various crises which await a state, it must severely feel the effects of any such deficiency. The extreme embarrassments of the United States during the late War, from an incapacity of supplying themselves, are still matter of keen recollection: A future war might be expected again to exemplify the mischiefs and dangers of a situation to which that incapacity is still in too great a degree applicable, unless changed by timely and vigorous exertion. To effect this change, as fast as shall be prudent, merits all the attention and all the Zeal of our Public Councils; it is the next great work to be accomplished.

The want of a Navy, to protect our external commerce, as long as it shall Continue, must render it a peculiarly precarious reliance, for the supply of essential articles, and must serve to strengthen prodigiously the arguments in favour of manufactures.

In order to a better judgment of the Means proper to be resorted to by the United States, it will be of use to Advert to those which have been employed with success in other Countries. The principal of these are—

1) Protecting Duties—Or Duties On Those Foreign Articles Which Are The Rivals Of The Domestic Ones Intended To Be Encouraged

Duties of this nature evidently amount to a virtual bounty on the domestic fabrics since by enhancing the charges on foreign articles, they enable the National Manufacturers to undersell all their foreign Competitors. . . .

2) Prohibitions Of Rival Articles, Or Duties Equivalent To Prohibitions

This is another and an efficacious mean of encouraging national manufactures but in general it is only fit to be employed when a manufacture, has made such progress and is in so many hands as to insure a due competition, and an adequate supply on reasonable terms. . . .

3) Prohibitions Of The Exportation Of The Materials Of Manufactures

The desire of securing a cheap and plentiful supply for the national workmen, and where the article is either peculiar to the Country, or of peculiar quality there, the jealousy of enabling foreign workmen to rival those of the nation with its own Materials, are the leading motives to this species of regulation. It ought not to be affirmed, that it is in no instance proper, but it is, certainly one which ought to be adopted with great circumspection, and only in very plain Cases. . . .

4) Pecuniary Bounties

This has been found one of the efficacious means of encouraging manufactures, and, is in some views, the best. . . .

Bounties have not, like high protecting duties, a tendency to produce scarcity. An increase of price is not always the immediate, though, where the progress of a domestic Manufacture does not counteract a rise, it is commonly the ultimate effect of an additional duty. In the interval, between the laying of the duty and the proportional increase of price, it may discourage importation, by interfering with the profits to be expected from the sale of the article.

It cannot escape notice, that the duty upon the importation of an article can no otherwise aid the domestic production of it, than by giving the latter greater advantages in the home market. It can have no influence upon the advantageous sale of the article produced in foreign markets; no tendency, therefore, to promote its exportation.

The true way to conciliate these two interests is to lay a duty on foreign *manufactures* of the material, the growth of which is desired to be encouraged, and to apply the produce of that duty, by way of bounty, either upon the production of the material itself or upon its manufacture at home, or upon both. . . .

5) Premiums

These are of a nature allied to bounties, though distinguishable from them in some important features.

Bounties are applicable to the whole quantity of an article produced, or manufactured, or exported, and involve a correspondent expense. Premiums serve to reward some particular excellence or superiority, some extraordinary exertion or skill, and are dispensed only in a small number of cases. . . .

6) The Exemption Of The Materials Of Manufactures From Duty

The policy of that Exemption as a general rule particularly in reference to new Establishments is obvious. It can hardly ever be advisable to add the obstructions of fiscal burthens to the difficulties which naturally embarrass a new manufacture; and where it is matured and in condition to become an object of revenue it is generally speaking better that the fabric than the Material should be the subject of Taxation. . . .

7) Drawbacks Of The Duties Which Are Imposed On The Materials Of Manufactures

It has already been observed as a general rule that duties on those materials ought with certain exceptions, to be forborne. Of these exceptions, three cases occur, which may serve as examples—one, where the material is itself an object of general or extensive consumption, and a fit and productive source of revenue: Another, where a manufacture of a simpler kind, the competition of which with a like domestic article is desired to be restrained, partakes of the Nature of a raw material, from being capable, by a further process to be converted into a manufacture of a different kind, the introduction or growth of which is desired to be encouraged; a third where the Material itself is a production of the country, and in sufficient abundance to furnish a cheap and plentiful supply to the national Manufacturers.

8) The Encouragement Of New Inventions And Discoveries, At Home, And Of The Introduction Into The United States Of Such As May Have Been Made In Other Countries; Particularly Those Which Relate To Machinery

. . . . The usual means of that encouragement are pecuniary rewards, and, for a time, exclusive privileges. The first must be employed according to the occasion, and the utility of the invention, or discovery. For the last, so far as respects "authors and inventors," provision has been made by Law. . . .

It is customary with manufacturing nations to prohibit, under severe penalties, the exportation of implements and machines, which they have either invented or improved. There are already objects for a similar regulation in the United States; and others may be expected to occur from time to time. The adoption of it seems to be dictated by the principle of reciprocity. Greater liberality, in such respects, might better comport with the general spirit of the country; but a selfish exclusive policy, in other quarters, will not always permit the free indulgence of a spirit which would place us upon an unequal footing. As far as prohibitions tend to prevent foreign competitors from deriving the benefit of the improvements made at home, they tend to increase the advantages of those by whom they may have been introduced, and operate as an encouragement to exertion.

9) Judicious Regulations For The Inspection Of Manufactured Commodities

. . . . Contributing to prevent frauds upon consumers at home and exporters to foreign countries, to improve the quality & preserve the character of the national manufactures, it cannot fail to aid the expeditious and advantageous sale of them, and to serve as a guard against successful competition from other quarters. . . .

10) The Facilitating Of Pecuniary Remittances From Place To Place

—is a point of considerable moment to trade in general, and to manufactures in particular; by rendering more easy the purchase of raw materials and provisions and the payment for manufactured supplies. A general circulation of Bank paper, which is to be expected from the institution lately established will be a most valuable mean to this end. . . .

11) The Facilitating Of The Transportation Of Commodities

The great copiousness of the subject of this Report has insensibly led to a more lengthy preliminary discussion than was originally contemplated, or intended. It appeared proper to investigate principles, to consider objections, and to endeavour to establish the utility of the thing proposed to be encouraged, previous to a specification of the objects which might occur, as meriting or requiring encouragement, and of the measures, which might be proper, in respect to each. The first purpose having been fulfilled, it remains to pursue the second.

In the selection of objects, five circumstances seem entitled to particular attention, the capacity of the Country to furnish the raw material—the degree in which the nature of the manufacture admits of a substitute for manual labour in machinery—the facility of execution—the extensiveness of the uses, to which the article can be applied—its subserviency to other interests, particularly the great one of national defence. There are however objects, to which these circumstances are little applicable, which for some special reasons, may have a claim to encouragement.

In countries where there is great private wealth, much may be effected by the voluntary contributions of patriotic individuals; but in a community situated like that of the United States, the public purse must supply the deficiency of private resource. In what can it be so useful, as in prompting and improving the efforts of industry?

Political and Cosmopolitical Economy

FRIEDRICH LIST

Before Quesnay and the French economists there existed only a *practice* of political economy which was exercised by the State officials, administrators, and authors who wrote about matters of administration, occupied themselves exclusively with the agriculture, manufactures, commerce, and navigation of those countries to which they belonged, without analysing the causes of wealth, or taking at all into consideration the interests of the whole human race.

Quesnay (from whom the idea of universal free trade originated) was the first who extended his investigations to the whole human race, without taking into consideration the idea of the nation. He calls his work "Physiocratie, ou du Gouvernement le plus avantageux au Genre Humain," his demands being that we must imagine that *the merchants of all nations formed one commercial republic.* Quesnay undoubtedly speaks of cosmopolitical economy, i.e. of that science which teaches how the entire human race may attain prosperity; in opposition to political economy, or that science which limits its teaching to the inquiry how a *given nation* can obtain (under the existing conditions of the world) prosperity, civilisation, and power, by means of agriculture, industry, and commerce.

Adam Smith treats his doctrine in a similarly extended sense, by making it his task to indicate the cosmopolitical idea of the absolute freedom of the commerce of the whole world in spite of the gross mistakes made by the physiocrates against the very nature of things and against logic. Adam Smith

This is a reprint of the 1885 edition of Friedrich List's "Political and Cosmopolitical Economy," in *The National System of Political Economy* by Reprints of Economic Classics. (New York: Augustus M. Kelley, 1966.)

concerned himself as little as Quesnay did with true political economy, i.e. that policy which each separate nation had to obey in order to make progress in its economical conditions. He entitles his work, "The Nature and Causes of the Wealth of Nations" (i.e. of all nations of the whole human race). He speaks of the various systems of political economy in a separate part of his work solely for the purpose of demonstrating their non-efficiency, and of proving that "political" or *national* economy must be replaced by "cosmopolitical or world-wide economy." Although here and there he speaks of wars, this only occurs incidentally. The idea of a perpetual state of peace forms the foundation of all his arguments. Moreover . . . his investigations from the commencement are based upon the principle that "most of the State regulations for the promotion of public prosperity are unnecessary, and a nation in order to be transformed from the lowest state of barbarism into a state of the highest possible prosperity needs nothing but bearable taxation, fair administration of justice, and *peace.*" . . .

J.B. Say openly demands that we should imagine the existence of a *universal republic* in order to comprehend the idea of general free trade. This writer, whose efforts were mainly restricted to the formation of a system out of the materials which Adam Smith had brought to light, says explicitly in the sixth volume (p. 288) of his "Economie politique pratique:" "We may take into our consideration the economical interests of the family with the father at its head; the principles and observations referring thereto will constitute *private economy.* Those principles, however, which have reference to the interests of whole nations, whether in themselves or in relation to other nations, form *public economy* (l'économie publique). *Political Economy,* lastly, relates to the interests of all nations, to *human society in general.*"

It must be remarked here, that in the first place Say recognises the existence of a national economy or political economy, under the name "économie publique," but that he nowhere treats of the latter in his works; secondly, that he attributes the name *political* economy to a doctrine which is evidently of *cosmopolitical* nature; and that in this doctrine he invariably merely speaks of an economy which has for its sole object the interests of the whole human society, without regard to the separate interests of distinct nations.

This substitution of terms might be passed over if Say, after having explained what he calls political economy (which, however, is nothing else but cosmopolitical or world-wide economy, or economy of the whole human race), had acquainted us with the principles of the doctrine which he calls "économie publique," which however is, properly speaking, nothing else but the economy of given nations, or true political economy.

In defining and developing this doctrine he could scarcely forbear to proceed from the idea and the nature of the nation, and to show what material modifications the "economy of the whole human race" must undergo by the fact that at present that race is still separated into distinct nationalities each held together by common powers and interests, and distinct from other societies of the same kind which in the exercise of their natural liberty are opposed to one another. However, by giving his cosmopolitical economy the name *political,* he dispenses with this explanation, effects by means of a transposition of terms also a transposition of meaning, and thereby masks a series of the gravest theoretical errors.

All later writers have participated in this error. Sismondi also calls political economy explicitly, "La science qui se charge du bonheur de l'espéce humaine." Adam Smith and his followers teach us from this mainly nothing more than what Quesnay and his followers had taught us already, for the article of the "Revue Méthodique" treating of the physiocratic school states, in almost the same words: *"The well-being of the individual is dependent altogether on the well-being of the whole human race."*

The first of the North American advocates of free trade, as understood by Adam Smith—Thomas Cooper, President of Columbia College—denies even the existence of nationality; he calls the nation "a grammatical invention," created only to save periphrases, a nonentity, which has no actual existence save in the heads of politicians. Cooper is moreover perfectly consistent with respect to this, in fact much more consistent than his predecessors and instructors, for it is evident that as soon as the existence of nations with their distinct nature and interests is recognised, it becomes necessary to modify the economy of human society in accordance with these special interests, and that if Cooper intended to represent these modifications as errors, it was very wise on his part from the beginning to disown the very existence of nations.

For our own part, we are far from rejecting the theory of *cosmopolitical* economy, as it has been perfected by the prevailing school; we are, however, of opinion that political economy, or as Say calls it "économie publique," should also be developed scientifically, and that it is always better to call things by their proper names than to give them significations which stand opposed to the true import of words.

If we wish to remain true to the laws of logic and of the nature of things, we must set the economy of individuals against the economy of societies, and discriminate in respect to the latter between true political or national economy (which, emanating from the idea and nature of the nation, teaches how a given *nation* in the present state of the world and its own special national relations can maintain and improve its economical conditions) and cosmopolitical economy, which originates in the assumption that all nations of the earth form but one society living in a perpetual state of peace.

If, as the prevailing school requires, we assume a universal union or confederation of all nations as the guarantee for an everlasting peace, the principle of international free trade seems to be perfectly justified. The less every individual is restrained in pursuing his own individual prosperity, the greater the number and wealth of those with whom he has free intercourse, the greater the area over which his individual activity can exercise itself, the easier it will be for him to utilise for the increase of his prosperity the properties given him by nature, the knowledge and talents which he has acquired, and the forces of nature placed at his disposal. As with separate individuals, so is it also the case with individual communities, provinces, and countries. A simpleton only could maintain that a union for free commercial intercourse between themselves is not as advantageous to the different states included in the United States of North America, to the various departments of France, and to the various German allied states, as would be their separation by internal provincial customs tariffs.

In the union of the three kingdoms of Great Britain and Ireland the world witnesses a great and irrefragable example of the immeasurable efficacy of free trade between united nations. Let us only suppose all other nations of the earth to be united in a similar manner, and the most vivid imagination will not be able to picture to itself the sum of prosperity and good fortune which the whole human race would thereby acquire.

A true principle, therefore, underlies the system of the popular school, but a principle which must be recognised and applied by science if its design to enlighten practice is to be fulfilled, an idea which practice cannot ignore without getting astray; only the school has omittted to take into consideration the nature of nationalities and their special interests and conditions, and to bring these into accord with the idea of universal union and an everlasting peace.

The popular school has assumed as being actually in existence a state of things which has yet to come into existence. It assumes the existence of a universal union and a state of perpetual peace, and deduces therefrom the great benefits of free trade. In this manner it confounds effects with causes. Among the provinces and states which are already politically united, there exists a state of perpetual peace; from this political union originates their commercial union, and it is in consequence of the perpetual peace thus maintained that the commercial union has become so beneficial to them. All examples which history can show are those in which the political union has led the way, and the commercial union has followed. Not a single instance can be adduced in which the latter has taken the lead, and the former has grown up from it. That, however, under the existing conditions of the world, the result of general free trade would not be a universal republic, but, on the contrary, a universal subjection of the less advanced nations to the supremacy of the predominant manufacturing, commercial, and naval power, is a conclusion for which the reasons are very strong and, according to our views, irrefragable. A universal republic (in the sense of Henry IV, and of the Abbé St. Pierre), i.e. a union of the nations of the earth whereby they recognise the same conditions of right among themselves and renounce self-redress, can only be realised if a large number of nationalities attain to as nearly the same degree as possible of industry and civilisation, political cultivation, and power. Only with the gradual formation of this union can free trade be developed, only as a result of this union can it confer on all nations the same great advantages which are now experienced by those provinces and states which are politically united. The system of protection, inasmuch as it forms the only means of placing those nations which are far behind in civilisation on equal terms with the one predominating nation (which, however, never received at the hands of Nature a perpetual right to a monopoly of manufacture, but which merely gained an advance over others in point of time), the system of protection regarded from this point of view appears to be the most efficient means of furthering the final union of nations, and hence also of promoting true freedom of trade. And national economy appears from this point of view to be that science which, correctly appreciating the existing interests and the individual circumstances of nations, teaches how *every separate nation* can be raised to that stage of industrial development in which union with other nations equally well developed, and consequently freedom of trade, can become possible and useful to it.

The popular school, however, has mixed up both doctrines with one another; it has fallen into the grave error of judging of the conditions of nations according to purely cosmopolitical principles, and of ignoring from merely political reasons the cosmopolitical tendency of the productive powers.

Only by ignoring the cosmopolitical tendency of the productive powers could Malthus be led into the error of desiring to restrict the increase of population, or Chalmers and Torrens maintain more recently the strange idea that augmentation of capital and unrestricted production are evils the restriction of which the welfare of the community imperatively demands, or Sismondi declare that manufactures are things injurious to the community. Their theory in this case resembles Saturn, who devours his own children— the same theory which allows that from the increase of population, of capital and machinery, division of labour takes place, and explains from this the welfare of society, finally considers these forces as monsters which threaten the prosperity of nations, because it merely regards the present conditions of individual nations, and does not take into consideration the conditions of the whole globe and the future progress of mankind.

It is not true that population increases in a larger proportion than production of the means of subsistence; it is at least foolish to assume such disproportion, or to attempt to prove it by artificial calculations or sophistical arguments, so long as on the glove a mass of natural forces still lies inert by means of which ten times or perhaps a hundred times more people than are now living can be sustained. It is mere narrow-mindedness to consider the present extent of the productive forces as the test of how many persons could be supported on a given area of land. . . . The culture of the potato and of food-yielding plants, and the more recent improvements made in agriculture generally, have increased tenfold the productive powers of the human race for the creation of the means of subsistence. In the Middle Ages the yield of wheat of an acre of land in England was fourfold, to-day it is ten to twenty fold, and in addition to that five times more land is cultivated. In many European countries (the soil of which possesses the same natural fertility as that of England) the yield at present does not exceed fourfold. Who will venture to set further limits to the discoveries, inventions, and improvements of the human race? Agricultural chemistry is still in its infancy; who can tell that to-morrow, by means of a new invention or discovery, the produce of the soil may not be increased five or ten fold? . . .

If in a nation the population increases more than the production of the means of subsistence, if capital accumulates at length to such an extent as no longer to find investment, if machinery throws a number of operatives out of work and manufactured goods accumulate to a large excess, this merely proves, that nature will not allow industry, civilisation, wealth, and power to fall exclusively to the lot of a single nation, or that a large portion of the globe suitable for cultivation should be merely inhabited by wild animals, and that the largest portion of the human race should remain sunk in savagery, ignorance, and poverty.

We have shown into what errors the school has fallen by judging the productive forces of the human race from a political point of view; we have now also to point out the mistakes which it has committed by regarding the separate interests of nations from a cosmopolitical point of view.

If a confederation of all nations existed in reality, as is the case with the separate states constituting the Union of North America, the excess of population, talents, skilled abilities, and material capital would flow over from England to the Continental states, in a similar manner to that in which it travels from the eastern states of the American Union to the western, provided that in the Continental states the same security for persons and property, the same constitution and general laws prevailed, and that the English Government was made subject to the united will of the universal confederation. Under these suppositions there would be no better way of raising all these countries to the same stage of wealth and cultivation as England than free trade. This is the argument of the school. But how would it tally with the actual operation of free trade under the existing conditions of the world?

The Britons as an independent and separate nation would henceforth take their national interest as the sole guide of their policy. The Englishman, from predilection for his language, for his laws, regulations, and habits, would wherever it was possible devote his powers and his capital to develop his own native industry, for which the system of free trade, by extending the market for English manufactures over all countries, would offer him sufficient opportunity; he would not readily take a fancy to establish manufactures in France or Germany. All excess of capital in England would be at once devoted to trading with foreign parts of the world. If the Englishman took it into his head to emigrate, or to invest his capital elsewhere than in England, he would as he now does prefer those more distant countries where he would find already existing his language, his laws, and regulations, rather than the benighted countries of the Continent. All England would thus be developed into one immense manufacturing city. Asia, Africa, and Australia would be civilised by England, and covered with new states modelled after the English fashion. In time a world of English states would be formed, under the presidency of the mother state, in which the European Continental nations would be lost as unimportant, unproductive races. By this arrangement it would fall to the lot of France, together with Spain and Portugal, to supply this English world with the choicest wines, and to drink the bad ones herself: at most France might retain the manufacture of a little millinery. Germany would scarcely have more to supply this English world with than children's toys, wooden clocks, and philological writings, and sometimes also an auxiliary corps, who might sacrifice themselves to pine away in the deserts of Asia or Africa, for the sake of extending the manufacturing and commercial supremacy, the literature and language of England. It would not require many centuries before people in this English world would think and speak of the Germans and French in the same tone as we speak at present of the Asiatic nations.

True political science, however, regards such a result of universal free trade as a very unnatural one; it will argue that had universal free trade been introduced at the time of the Hanseatic League, the German nationality instead of the English would have secured an advance in commerce and manufacture over all other countries.

It would be most unjust, even on cosmopolitical grounds, now to resign to the English all the wealth and power of the earth, merely because by them the political system of commerce was first established and the cosmopolitical principle for the most part ignored. In order to allow freedom of trade to operate naturally, the less advanced nations must first be raised by artificial measures to that stage of cultivation to which the English nation has been artificially elevated. In order that, through that cosmopolitical tendency of the powers of production to which we have alluded, the more distant parts of the world may not be benefited and enriched before the neighboring European countries, those nations which feel themselves to be capable, owing to their moral, intellectual, social, and political circumstances, of developing a manufacturing power of their own must adopt the system of protection as the most effectual means for this purpose. The effects of this system for the purpose in view are of two kinds; in the first place, by gradually excluding foreign manufactured articles from our markets, a surplus would be occasioned in foreign nations, of workmen, talents, and capital, which must seek employment abroad; and secondly, by the premium which our system of protection would offer to the immigration into our country of workmen, talents, and capital, that excess of productive power would be induced to find employment with us, instead of emigrating to distant parts of the world and to colonies. Political science refers to history, and inquires whether England has not in former times drawn from Germany, Italy, Holland, France, Spain, and Portugal by these means a mass of productive power. She asks: Why does the cosmopolitical school, when it pretends to weigh in the balance the advantages and the disadvantages of the system of protection, utterly ignore this great and remarkable instance of the results of that system?

2

Classic Liberalism

Liberal theorists redefined political economy. By focusing on the logic and consequences of economic action, they undermined mercantilist theory. The early liberals did not wholly separate economics from politics, as is often done by contemporary neoclassical economists. They understood that economic policy was infused with particularistic interests. Liberals argued, however, that economic development and transformation could not be attained if entrenched political interests continued to subvert market forces. Moreover, if markets were allowed to function freely, then the broader political interest of the many, if not the entire community, would be better served. Liberalism thus reversed the mercantilist formula and suggested that economic reason could ultimately resolve conflicting political interests.

Adam Smith (1723–1790), drawing upon emergent liberal themes as well as his own insights, led the way in attacking mercantilist orthodoxy and creating a new approach to political economy. Although mercantilist theory was losing its vitality, it still dominated economic policies of his day—and would continue to for long after. Smith therefore had to begin with a thoroughgoing critique of mercantilism. He went right to the heart of the matter, analyzing the nature of wealth. For Smith, wealth is not merely a function of the amount of gold and silver in state coffers. Rather, wealth flows from the general productive capacity of an economy. National wealth, and with it power, increases only through comprehensive economic growth. To gain this end, the state need not intervene in the market. Productive gains are best secured by rational egoists freely pursuing individual material interests. Internationally, Smith suggests that such productive gains are not necessarily zero–sum. If the inherent productivity rewards of a free market are allowed to spread around the world via free trade, the wealth of all nations is ensured.

Smith further elaborates the rationale for free trade. Domestically, efficient production and wealth rest upon a division of labor and economic interdependence. These principles also apply to the world economy: "what is prudent in the conduct of every private family, can scarce be folly in that of a great kingdom." Just as government agents cannot effectively manage the myriad economic interrelations of the national economy, so too, states should not interfere in world markets. If placed in an unregulated international economy, each nation would discover a productive niche, an absolute or comparative advantage. The division of labor can, and should, be replicated globally. Smith is not Panglossian in outlook, however. He realizes that vested interests, especially merchants, benefit from protectionism and will fight to retain their advantages. But these specific interests should not be confused with the broader national interest, which is better served by the prosperity flowing from free trade.

Smith does not call for unilateral and unconditional free trade. He offers four qualifications to his argument. First, free trade may be limited by the exigencies of national security. Smith sympathizes with the Navigation Acts in this regard: "the wisest of all commercial regulations of England." Second, duties should be imposed on foreign goods equal to domestic taxes on national products in order to avoid offering undue advantage to imports. Third, duties can be imposed as retaliation for unfair restrictions in foreign markets but only as a means of forcing other states to rescind their duties. Retaliation in and of itself is more harmful than good. Finally, free trade should be phased in gradually so that domestic industry and labor have time to make necessary adjustments to heightened international competition. Smith thus tempers his argument with an understanding of prevailing mercantilist practice and domestic political realities.

David Ricardo (1772–1823) builds upon, and moves away from, Smith. He, too, sees political interests behind existing economic policies. Yet where Smith argues against mercantilist businessmen, Ricardo regards landed interests as more subversive to economic progress. His reasoning is bound up with his concepts of rent, wages, and profits. Accepting Thomas Malthus's notion of finite resources (one of the few things on which these two friends agreed), Ricardo foresaw rising food prices that would push up rents accruing to landlords. Higher food prices would also cause wages to rise, undermining profits for the fledgling bourgeoisie. The landlord's gain was the industrialist's loss. Without adequate profits, businessmen could not invest and economic development would suffer. World markets offer a means of overcoming the agricultural obstacle to development. Ricardo, therefore, opens the chapter on foreign trade in his *Principles of Political Economy and Taxation* with a discussion of the relationship of trade to domestic profits and wages. Foreign trade is nationally beneficial only insofar as it contributes to lower wages, which in turn will allow profits to rise.

To gain the most from foreign trade, however, nations should concentrate on comparative advantage. Ricardo's famous example of British cloth and Portuguese wine makes a powerful case for economic specialization based on comparative costs and open world markets. He is careful to consider this argument in light of international payments imbalances. Monetary forces, he finds, do not endanger comparative advantage because of the self-correcting specie flow mechanism (the forerunner of the quantity theory of money). If there is a flaw in this enduring argument, it may be the assumption of immobile capital and labor, conditions that have changed in the era of the transnational corporation and modern transportation.

For Ricardo, international economics is not like domestic economics. A national economy is constrained by finite resources and conflicting interests; the world economy offers untapped avenues of growth and expansion. He does not ponder global exhaustion, a remote possibility in his day. In sum, although they differ in methodology—Ricardo's abstract theory contrasting sharply with Smith's narrative style—they agree on the positive results of free trade.

Of the Principle of the Commercial or Mercantile System

ADAM SMITH

That wealth consists in money, or in gold and silver, is a popular notion which naturally arises from the double function of money, as the instrument of commerce, and as the measure of value. In consequence of its being the instrument of commerce, when we have money we can more readily obtain whatever else we have occasion for, than by means of any other commodity. The great affair, we always find, is to get money. When that is obtained, there is no difficulty in making any subsequent purchase. In consequence of its being the measure of value, we estimate that of all other commodities by the quantity of money which they will exchange for. We say of a rich man that he is worth a great deal, and of a poor man that he is worth very little money. A frugal man, or a man eager to be rich, is said to love money; and a careless, a generous, or a profuse man, is said to be indifferent about it. To grow rich is to get money; and wealth and money, in short, are, in common language, considered as in every respect synonymous.

A rich country, in the same manner as a rich man, is supposed to be a country abounding in money; and to heap up gold and silver in any country is supposed to be the readiest way to enrich it. . . .

Mr. Locke remarks a distinction between money and other moveable goods. All other moveable goods, he says, are of so consumable a nature that the wealth which consists in them cannot be much depended on, and a nation which abounds in them one year may, without any exportation, but merely by their own waste and extravagance, be in great want of them the next. Money,

Reprinted from Adam Smith, "Of the Principle of the Commercial or Mercantile System," *The Wealth of Nations*. (New York: Modern Library, 1937.)

on the contrary, is a steady friend, which, though it may travel about from hand to hand, yet if it can be kept from going out of the country, is not very liable to be wasted and consumed. Gold and silver, therefore, are, according to him, the most solid and substantial part of the moveable wealth of a nation, and to multiply those metals ought, he thinks, upon that account, to be the great object of its political economy.

Others admit that if a nation could be separated from all the world, it would be of no consequence how much, or how little money circulated in it. The consumable goods which were circulated by means of this money, would only be exchanged for a greater or a smaller number of pieces; but the real wealth or poverty of the country, they allow, would depend altogether upon the abundance or scarcity of those consumable goods. But it is otherwise, they think, with countries which have connections with foreign nations, and which are obliged to carry on foreign wars, and to maintain fleets and armies in distant countries. This, they say, cannot be done, but by sending abroad money to pay them with; and a nation cannot send much money abroad, unless it has a good deal at home. Every such nation, therefore, must endeavour in time of peace to accumulate gold and silver, that, when occasion requires, it may have wherewithal to carry on foreign wars.

In consequence of these popular notions, all the different nations of Europe have studied, though to little purpose, every possible means of accumulating gold and silver in their respective countries. Spain and Portugal, the proprietors of the principal mines which supply Europe with those metals, have either prohibited their exportation under the severest penalties, or subjected it to a considerable duty. The like prohibition seems anciently to have made a part of the policy of most other European nations. . . .

When those countries became commercial, the merchants found this prohibition, upon many occasions, extremely inconvenient. They could frequently buy more advantageously with gold and silver than with any other commodity, the foreign goods which they wanted, either to import into their own, or to carry to some other foreign country. They remonstrated, therefore, against this prohibition as hurtful to trade.

They represented, first, that the exportation of gold and silver in order to purchase foreign goods, did not always diminish the quantity of those metals in the kingdom. That, on the contrary, it might frequently increase that quantity; because, if the consumption of foreign goods was not thereby increased in the country, those goods might be re-exported to foreign countries, and, being there sold for a large profit, might bring back much more treasure than was originally sent out to purchase them. . . .

They represented, secondly, that this prohibition could not hinder the exportation of gold and silver, which, on account of the smallness of their bulk in proportion to their value, could easily be smuggled abroad. That this exportation could only be prevented by a proper attention to, what they called, the balance of trade. That when the country exported to a greater value than it imported, a balance became due to it from foreign nations, which was necessarily paid to it in gold and silver, and thereby increased the quantity of those metals in the kingdom. But that when it imported to a greater value than it exported, a contrary balance became due to foreign nations, which was necessarily paid to them in the same manner, and thereby diminished that quantity. . . .

Those arguments were partly solid and partly sophistical. They were solid so far as they asserted that the exportation of gold and silver in trade might frequently be advantageous to the country. They were solid too, in asserting that no prohibition could prevent their exportation, when private people found any advantage in exporting them. But they were sophistical in supposing, that either to preserve or to augment the quantity of those metals required more the attention of government, than to preserve or to augment the quantity of any other useful commodities, which the freedom of trade, without any such attention, never fails to supply in the proper quantity. They were sophistical too, perhaps, in asserting that the high price of exchange necessarily increased, what they called, the unfavourable balance of trade, or occasioned the exportation of a greater quantity of gold and silver. . . . The high price of exchange . . . would naturally dispose the merchants to endeavour to make their exports nearly balance their imports, in order that they might have this high exchange to pay upon as small a sum as possible. The high price of exchange, besides, must necessarily have operated as a tax, in raising the price of foreign goods, and thereby diminishing their consumption. It would tend, therefore, not to increase, but to diminish, what they called, the unfavourable balance of trade, and consequently the exportation of gold and silver.

Such as they were, however, those arguments convinced the people to whom they were addressed. They were addressed by merchants to parliaments, and to the councils of princes, to nobles, and to country gentlemen; by those who were supposed to understand trade, to those who were conscious to themselves that they knew nothing about the matter. That foreign trade enriched the country, experience demonstrated to the nobles and country gentlemen, as well as to the merchants; but how, or in what manner, none of them well knew. The merchants knew perfectly in what manner it enriched themselves. It was their business to know it. But to know in what manner it enriched the country, was no part of their business. This subject never came into their consideration, but when they had occasion to apply to their country for some change in the laws relating to foreign trade. It then became necessary to say something about the beneficial effects of foreign trade, and the manner in which those effects were obstructed by the laws as they then stood. To the judges who were to decide the business, it appeared a most satisfactory account of the matter, when they were told that foreign trade brought money into the country, but that the laws in question hindered it from bringing so much as it otherwise would do. Those arguments therefore produced the wished-for effect. . . . The attention of government was turned away from guarding against the exportation of gold and silver, to watch over the balance of trade, as the only cause which could occasion any augmentation or diminution of those metals. From one fruitless care it was turned away to another care much more intricate, much more embarrassing, and just equally fruitless. . . .

A country that has no mines of its own must undoubtedly draw its gold and silver from foreign countries, in the same manner as one that has no vineyards of its own must draw its wines. It does not seem necessary, however, that the attention of government should be more turned towards the one than towards the other object. A country that has wherewithal to buy wine, will always get the wine which it has occasion for; and a country that has wherewithal to buy gold and silver, will never be in want of those metals.

They are to be bought for a certain price like all other commodities, and as they are the price of all other commodities, so all other commodities are the price of those metals. We trust with perfect security that the freedom of trade, without any attention of government, will always supply us with the wine which we have occasion for: and we may trust with equal security that it will always supply us with all the gold and silver which we can afford to purchase or to employ, either in circulating our commodities, or in other uses.

The quantity of every commodity which human industry can either purchase or produce, naturally regulates itself in every country according to the effectual demand, or according to the demand of those who are willing to pay the whole rent, labour and profits which must be paid in order to prepare and bring it to market. But no commodities regulate themselves more easily or more exactly according to this effectual demand than gold and silver; because, on account of the small bulk and great value of those metals, no commodities can be more easily transported from one place to another, from the places where they are cheap, to those where they are dear, from the places where they exceed, to those where they fall short of this effectual demand. . . .

When the quantity of gold and silver imported into any country exceeds the effectual demand, no vigilance of government can prevent their exportation. . . . If, on the contrary, in any particular country their quantity fell short of the effectual demand, so as to raise their price above that of the neighboring countries, the government would have no occasion to take any pains to import them. If it were even to take pains to prevent their importation it would not be able to effectuate it. . . .

If, not withstanding all this, gold and silver should at any time fall short in a country which has wherewithal to purchase them, there are more expedients for supplying their place, than that of almost any other commodity. If the materials of manufacture are wanted, industry must stop. If provisions are wanted, the people must starve. But if money is wanted, barter will supply its place, though with a good deal of inconveniency. Buying and selling upon credit, and the different dealers compensating their credits with one another, once a month or once a year, will supply it with less inconveniency. A well-regulated paper money will supply it, not only without any inconveniency, but, in some cases, with some advantages. Upon every account, therefore, the attention of government never was so unnecessarily employed, as when directed to watch over the preservation or increase of the quantity of money in any country.

It would be too ridiculous to go about seriously to prove, that wealth does not consist in money, or in gold and silver; but in what money purchases, and is valuable only for purchasing. Money, no doubt, makes always a part of the national capital; but . . . it generally makes but a small part, and always the most unprofitable part of it.

It is not because wealth consists more essentially in money than in goods, that the merchant finds it generally more easy to buy goods with money, than to buy money with goods; but because money is the known and established instrument of commerce, for which every thing is readily given in exchange, but which is not always with equal readiness to be got in exchange for every thing. . . . Goods can serve many other purposes besides purchasing money, but money can serve no other purpose besides purchasing goods. Money, therefore, necessarily runs after goods, but goods do not always or necessarily run after money. The man who buys, does not always mean to sell again, but frequently to use or to consume; whereas he who sells, always means to buy again. The one may frequently have done the whole, but the other can never have done more than the one-half of his business. It is not for its own sake that men desire money, but for the sake of what they can purchase with it.

. . . Gold and silver, whether in the shape of coin or plate, are utensils, it must be remembered, as much as the furniture of the kitchen. Increase the use for them, increase the consumable commodities which are to be circulated, managed, and prepared by means of them, and you will infallibly increase the quantity; but if you attempt, by extraordinary means, to increase the quantity, you will as infallibly diminish the use and even the quantity too, which in those metals can never be greater than what the use requires. Were they ever to be accumulated beyond this quantity, their transportation is so easy, and the loss which attends their lying idle and unemployed so great, that no law could prevent their being immediately sent out of the country.

It is not always necessary to accumulate gold and silver, in order to enable a country to carry on foreign wars, and to maintain fleets and armies in distant countries. Fleets and armies are maintained, not with gold and silver, but with consumable goods. The nation which, from the annual produce of its domestic industry, from the annual revenue arising out of its lands, labour, and consumable stock, has wherewithal to purchase those consumable goods in distant countries, can maintain foreign wars there.

A nation may purchase the pay and provisions of an army in a distnat country three different ways; by sending abroad either, first, some part of its accumulated gold and silver; or secondly, some part of the annual produce of its manufactures; or last of all, some part of its annual rude produce.

The gold and silver which can properly be considered as accumulated or stored up in any country, may be distinguished into three parts; first, the circulating money; secondly, the plate of private families; and last of all, the money which may have been collected by many years parsimony, and laid up in the treasury of the prince.

The funds which maintained the foreign wars of the present century, the most expensive perhaps which history records, seem to have had little dependency upon the exportation either of the circulating money, or of the plate of private families, or of the treasure of the prince. . . .

The enormous expense of the late war, therefore, must have been chiefly defrayed, not by the exportation of gold and silver, but by that of British commodities of some kind or other. . . .

The commodities most proper for being transported to distnat countries, in order to purchase there, either the pay and provisions of an army, or some part of the money of the mercantile republic to be employed in purchasing them, seem to be the finer and more improved manufactures; such as contain a great value in a small bulk, and can, therefore, be exported to a great distance at little expense. A country whose industry produces a great annual surplus of such manufactures, which are usually exported to foreign countries, may carry on for many years a very expensive foreign war, without either exporting any considerable quantity of gold and silver, or even having any such quantity to export. . . .

No foreign war of great expense or duration could conveniently be carried on by the exportation of the rude produce of the soil. The expense of sending such a quantity of it to a foreign country as might purchase the pay and provisions of an army, would be too great. Few countries too produce much more rude produce than what is sufficient for the subsistence of their own inhabitants. To send abroad any great quantity of it, therefore, would be to send abroad a part of the necessary subsistence of the people. It is otherwise with the exportation of manufactures. The maintenance of the people employed in them is kept at home, and only the surplus part of their work is exported. . . .

The importation of gold and silver is not the principal, much less the sole benefit which a nation derives from its foreign trade. Between whatever places foreign trade is carried on, they all of them derive two distinct benefits from it. It carries out that surplus part of the produce of their land and labour for which there is no demand among them, and brings back in return for it something else for which there is a demand. It gives a value to their superfluities, by exchanging them for something else, which may satisfy a part of their wants, and increase their enjoyments. By means of it, the narrowness of the home market does not hinder the division of labour in any particular branch of art or manufacture from being carried to the highest perfection. By opening a more extensive market for whatever part of the produce of their labour may exceed the home consumption, it encourages them to improve its productive powers, and to augment its annual produce to the utmost, and thereby to increase the real revenue and wealth of the society. These great and important services foreign trade is continually occupied in performing, to all the different countries between which it is carried on. They all derive great benefit from it, though that in which the merchant resides generally derives the greatest, as he is generally more employed in supplying the wants, and carrying out the superfluities of his own, than of any other particular country. To import the gold and silver which may be wanted, into the countries which have no mines, is, no doubt, a part of the business of foreign commerce. It is, however, a most insignificant part of it. A country which carried on foreign trade merely upon this account, could scarce have occasion to freight a ship in a century.

I thought it necessary, though at the hazard of being tedious, to examine at full length this popular notion that wealth consists in money, or in gold and silver. Money in common language, as I have already observed, frequently signifies wealth; and this ambiguity of expression has rendered this popular notion so familiar to us, that even they, who are convinced of its absurdity, are very apt to forget their own principles, and in the course of their reasonings to take it for granted as a certain and undeniable truth. Some of the best English writers upon commerce set out with observing, that the wealth of a country consists, not in its gold and silver only, but in its lands, houses, and consumable goods of all different kinds. In the course of their reasonings, however, the lands, houses, and consumable goods seem to slip out of their memory, and the strain of their argument frequently supposes that all wealth consists in gold and silver, and that to multiply those metals is the great object of national industry and commerce.

The two principles being established, however, that wealth consisted in gold and silver, and that those metals could be brought into a country which had no mines only by the balance of trade, or by exporting to a greater value than it imported; it necessarily became the great object of political economy to diminish as much as possible the importation of foreign goods for home consumption, and to increase as much as possible the exportation of the produce of domestic industry. Its two great engines for enriching the country, therefore, were restraints upon importation, and encouragements to exportation.

Of Restraints Upon the Importation from Foreign Countries of Such Goods as Can Be Produced at Home

ADAM SMITH

By restraining, either by high duties, or by absolute prohibitions, the importation of such goods from foreign countries as can be produced at home, the monopoly of the home market is more or less secured to the domestic industry employed in producing them. . . .

That this monopoly of the home-market frequently gives great encouragement to that particular species of industry which enjoys it, and frequently turns towards that employment a greater share of both the labour and stock of the society than would otherwise have gone to it, cannot be doubted. But whether it tends either to increase the general industry of the society, or to give it the most advantageous direction, is not, perhaps, altogether so evident.

The general industry of the society never can exceed what the capital of the society can employ. As the number of workmen that can be kept in employment by any particular person must bear a certain proportion to his capital, so the number of those that can be continually employed by all the members of a great society, must bear a certain proportion to the whole capital of that society, and never can exceed that proportion. No regulation of commerce can increase the quantity of industry in any society beyond what its capital can maintain. It can only divert a part of it into a direction into which it might not otherwise have gone; and it is by no means certain that this artificial direction is likely to be more advantageous to the society than that into which it would have gone of its own accord.

Every individual is continually exerting himself to find out the most advantageous employment for whatever capital he can command. It is his own advantage, indeed, and not that of the society, which he has in view. But the study of his own advantage naturally, or rather necessarily leads him to prefer that employment which is most advantageous to the society.

First, every individual endeavours to employ his capital as near home as he can, and consequently as much as he can in the support of domestic industry; provided always that he can thereby obtain the ordinary, or not a great deal less than the ordinary profits of stock.

Thus, upon equal or nearly equal profits, every wholesale merchant naturally prefers the home-trade to the foreign trade of consumption, and the foreign trade of consumption to the carrying trade. In the home-trade his capital is never so long out of his sight as it frequently is in the foreign trade of consumption. He can know better the character and situation of the persons whom he trusts, and if he should happen to be deceived, he knows better the laws of the country from which he must seek redress. In the carrying trade, the capital of the merchant is, as it were, divided between two foreign countries, and no part of it is ever necessarily brought home, or placed under his own immediate view and command. . . . But a capital employed in the home-trade, it has already been shown, necessarily puts into motion a greater quantity of domestic industry, and gives revenue and employment to a greater number of the inhabitants of the country, than an equal capital employed in the foreign trade of consumption: and one employed in the foreign trade of consumption has the same advantage over an equal capital employed in the carrying trade. Upon equal, or only nearly equal profits, therefore, every individual naturally inclines to employ his capital in the manner in which it is likely to afford the greatest support to domestic industry, and to give revenue and employment to the greatest number of people of his own country.

Secondly, every individual who employs his capital in the support of domestic industry, necessarily endeavours so to direct that industry, that its produce may be of the greatest possible value.

The produce of industry is what it adds to the subject or materials upon which it is employed. In proportion as the value of this produce is great or small, so will likewise be the profits of the employer. But it is only for the sake of profit that any man employs a capital in the support of industry; and he will always, therefore, endeavour to employ it in the support of that industry of which the produce is likely to be of the greatest value, or to exchange for the greatest quantity either of money or of other goods.

But the annual revenue of every society is always precisely equal to the exchangeable value of the whole annual produce of its industry, or rather is precisely the same thing with that exchangeable value. As every individual, therefore, endeavours as much as he can both to employ his capital in the support of domestic industry, and so to direct that industry that its produce may be of the greatest value; every individual necessarily labours to render the annual revenue of the society as great as he can. He generally, indeed, neither intends to promote the public interest, nor knows how much he is promoting it. By preferring the support of domestic to that of foreign industry, he intends only his own security; and by directing that industry in such a manner as its produce may be of the greatest value, he intends only his own gain, and he is in this, as in many other cases, led by an invisible hand to promote an end which was no part of his intention. Nor is it always the worse for the society that it was no part of it. By pursuing his own interest he frequently promotes that of the society more effectually than when he really intends to promote it. I have never known much good done by those who affected to trade for the public good. It is an affectation, indeed, not very common among merchants, and very few words need be employed in dissuading them from it.

What is the species of domestic industry which his capital can employ, and of which the produce is likely to be of the greatest value, every individual, it is evident, can, in his local situation, judge much better than any statesman or lawgiver can do for him. The statesman, who should attempt to direct private people in what manner they ought to employ their capitals, would not only load himself with a most unnecessary attention, but assume an authority which could safely be trusted, not only to no single person, but to no council or senate whatever, and which would nowhere be so dangerous as in the hands of a man who had folly and presumption enough to fancy himself fit to exercise it.

To give the monopoly of the home-market to the produce of domestic industry, in any particular art or manufacture, is in some measure to direct private people in what manner they ought to employ their capitals, and must, in almost all cases, be either a useless or a hurtful regulation. If the produce of domestic can be brought there as cheap as that of foreign industry, the regulation is evidently useless. If it cannot, it must generally be hurtful. It is the maxim of every prudent master of a family, never to attempt to make at home what it will cost him more to make than to buy. The taylor does not attempt to make his own shoes, but buys them of the shoemaker. The shoemaker does not attempt to make his own clothes, but employs a taylor. The farmer attempts to make neither the one nor the other, but employs those different artificers. All of them find it for their interest to employ their whole industry in a way in which they have some advantage over their neighbours, and to purchase with a part of its produce, or what is the same thing, with the price of a part of it, whatever else they have occasion for.

What is prudence in the conduct of every private family, can scarce be folly in that of a great kingdom. If a foreign country can supply us with a commodity cheaper than we ourselves can make it, better buy it of them with some part of the produce of our own industry, employed in a way in which we have some advantage. The general industry of the country, being always in proportion to the capital which employs it, will not thereby be diminished, no more than that of the above-mentioned artificers; but only left to find out the way in which it can be employed with the greatest advantage. It is certainly not employed to the greatest advantage, when it is thus directed towards an object which it can buy cheaper than it can make. The value of its annual produce is certainly more or less diminished, when it is thus turned away from producing commodities evidently of more value than the commodity which it is directed to produce. According to the supposition, that commodity could be purchased from foreign countries cheaper than it can be made at home. It could, therefore, have been purchased with a part only of the commodities, or, what is the same thing, with a part only of the price of the commodities, which the industry employed by an equal capital would have produced at home, had it been left to follow its natural course. The industry of the country, therefore, is thus turned away from a more, to a less advantageous employment, and the exchangeable value of its annual produce, instead of being increased, according to the intention of the lawgiver, must necessarily be diminished by every such regulation.

By means of such regulations, indeed, a particular manufacture may sometimes be acquired sooner than it could have been otherwise, and after a certain time may be made at home as cheap or cheaper than in the foreign country. But though the industry of the society may be thus carried with advantage into a particular channel sooner than it could have been otherwise, it will by no means follow that the sum total, either of its industry, or of its revenue, can ever be augmented by any such regulation. The industry of the society can augment only in proportion as its capital augments, and its capital can augment only in proportion to what can be gradually saved out of its revenue. But the immediate effect of every such regulation is to diminish its revenue, and what diminishes its revenue is certainly not very likely to augment its capital faster than it would have augmented of its own accord, had both capital and industry been left to find out their natural employments.

Though for want of such regulations the society should never acquire the proposed manufacture, it would not, upon that account, necessarily be the poorer in any one period of its duration. In every period of its duration its whole capital and industry might still have been employed, though upon different objects, in the manner that was most advantageous at the time. In every period its revenue might have been the greatest which its capital could afford, and both capital and revenue might have been augmented with the greatest possible rapidity.

The natural advantages which one country has over another in producing particular commodities are sometimes so great, that it is acknowledged by all the world to be in vain to struggle with them. . . . Whether the advantages which one country has over another, be natural or acquired, is in this respect of no consequence. As long as the one country has those advantages, and the other wants them, it will always be more advantageous for the latter, rather to buy of the former than to make. It is an acquired advantage only, which one artificer has over his neighbour, who exercises another trade; and yet they both find it more advantageous to buy of one another, than to make what does not belong to their particular trades.

There seem, however, to be two cases in which it will generally be advantageous to lay some burden upon foreign, for the encouragement of domestic industry.

The first is, when some particular sort of industry is necessary for the reference of the country. The defence of Great Britain, for example, depends very much upon the number of its sailors and shipping. The act of navigation, therefore, very properly endeavours to give the sailors and shipping of Great Britain the monopoly of the trade of their own country, in some cases, by absolute prohibitions, and in others by heavy burdens upon the shipping of foreign countries. . . .

The act of navigation is not favourable to foreign commerce, or to the growth of that opulence which can arise from it. The interest of a nation in its commercial relations to foreign nations is, like that of a merchant with regard to the different people with whom he deals, to buy as cheap and to sell as dear as possible. But it will be most likely to buy cheap, when by the most perfect freedom of trade it encourages all nations to bring to it the goods which it has occasion to purchase; and, for the same reason, it will be most likely to sell dear, when its markets are thus filled with the greatest number of buyers. The act of navigation, it is true, lays no burden upon foreign ships that come to export the produce of British industry. Even the ancient aliens duty, which used to be paid upon all goods exported as well as imported, has, by several subsequent acts, been taken off from the greater part of the articles of exportation. But if foreigners, either by prohibitions or high duties, are hindered from coming to sell, they cannot always afford to come to buy; because coming without a cargo, they must lose the freight from their own country to Great Britain. By diminishing the number of sellers, therefore, we necessarily diminish that of buyers, and are thus likely not only to buy foreign goods dearer, but to sell our own cheaper, than if there was a more perfect freedom of trade. As defence, however, is of much more importance than opulence, the act of navigation is, perhaps, the wisest of all the commercial regulations of England.

The second case, in which it will generally be advantageous to lay some burden upon foreign for the encouragement of domestic industry, is, when some tax is imposed at home upon the produce of the latter. In this case, it seems reasonable that an equal tax should be imposed upon the like produce of the former. This would not give the monopoly of the home market to domestic industry, nor turn towards a particular employment a greater share of the stock and labour of the country, than what would naturally go to it. It would only hinder any part of what would naturally go to it from being turned away by the tax, into a less natural direction, and would leave the competition between foreign and domestic industry, after the tax, as nearly as possible upon the same footing as before it. . . .

This second limitation of the freedom of trade according to some people should, upon some occasions, be extended much farther than to the precise foreign commodities which could come into competition with those which had been taxed at home. When the necessaries of life have been taxed in any country, it becomes proper, they pretend, to tax not only the like necessaries of life imported from other countries, but all sorts of foreign goods which can come into competition with any thing that is the produce of domestic industry. Subsistence, they say, becomes necessarily dearer in consequence of such taxes; and the price of labour must always rise with the price of the labourers subsistence. Every commodity, therefore, which is the produce of domestic industry, though not immediately taxed itself, becomes dearer in consequence of such taxes, because the labour which produces it becomes so. Such taxes, therefore, are really equivalent, they say, to a tax upon every particular commodity produced at home. In order to put domestic upon the same footing with foreign industry, therefore, it becomes necessary, they think, to lay some duty upon every foreign commodity, equal to this enhancement of the price of the home commodities with which it can come into competition.

Such taxes, when they have grown up to a certain height, are a curse equal to the barrenness of the earth and the inclemency of the heavens; and yet it is in the richest and most industrious countries that they have been most generally imposed. No other countries could support so great a disorder. As the strongest bodies only can live and enjoy health, under an unwholesome regimen; so the nations only, that in every sort of industry have the greatest natural and acquired advantages, can subsist and prosper under such taxes. Holland is the country in Europe in which they abound most, and which from peculiar circumstances continues to prosper, not by means of them, as has been most absurdly supposed, but in spite of them.

As there are two cases in which it will generally be advantageous to lay some burden upon foreign, for the encouragement of domestic industry; so there are two others in which it may sometimes be a matter of deliberation; in the one, how far it is proper to continue the free importation of certain foreign goods; and in the other, how far, or in what manner, it may be proper to restore that free importation after it has been for some time interrupted.

The case in which it may sometimes be a matter of deliberation how far it is proper to continue the free importation of certain foreign goods, is, when some foreign nation restrains by high duties or prohibitions the importation of some of our manufactures into their country. Revenge in this case naturally dictates retaliation, and that we should impose the like duties and prohibitions upon the importation of some or all of their manufactures into ours. Nations accordingly seldom fail to retaliate in this manner. . . .

There may be good policy in retaliations of this kind, when there is a probability that they will procure the repeal of the high duties or prohibitions complained of. The recovery of a great foreign market will generally more than compensate the transitory inconveniency of paying dearer during a short time for some sorts of goods . . . When there is no probability that any such repeal can be procured, it seems a bad method of compensating the injury done to certain classes of our people, to do another injury ourselves, not only to those classes, but to almost all the other classes of them. . . .

The case in which it may sometimes be a matter of deliberation, how far, or in what manner, it is proper to restore the free importation of foreign goods, after it has been for some time interrupted, is, when particular manufactures, by means of high duties or prohibitions upon all foreign goods which can come into competition with them, have been so far extended as to employ a great multitude of hands. Humanity may in this case require that the freedom of trade should be restored only by slow gradations, and with a good deal of reserve and circumspection. Were those high duties and prohibitions taken away all at once, cheaper foreign goods of the same kind might be poured so fast into the home market, as to deprive all at once many thousands of our people of their ordinary employment and means of subsistence. The disorder which this would occasion might no doubt be very considerable. It would in all probability, however, be much less than is commonly imagined. . . .

To expect, indeed, that the freedom of trade should ever be entirely restored in Great Britain, is as absurd as to expect that an Oceana or Utopia should ever be established in it. Not only the prejudices of the public, but what is much more unconquerable, the private interests of many individuals, irresistibly oppose it. Were the officers of the army to oppose with the same zeal and unanimity any reduction in the number of forces, with which master manufacturers set themselves against every law that is likely to increase the number of their rivals in the home market; were the former to animate their soldiers, in the same manner as the latter enflame their workmen, to attack with violence and outrage the proposers of any such regulation; to attempt to reduce the army would be as dangerous as it has now become to attempt to diminish in any respect the monopoly which our manufacturers have obtained against us. This monopoly has so much increased the number of some particular tribes of them, that, like an overgrown standing army, they have become formidable to the government, and upon many occasions intimidate the legislature. The member of parliament who supports every proposal for strengthening this monopoly, is sure to acquire not only the reputation of understanding trade, but great popularity and influence with an order of men whose numbers and wealth render them of great importance. If he opposes them, on the contrary, and still more if he has authority enough to be able to thwart them, neither the most acknowledged probity, nor the highest rank, nor the greatest public services, can protect him from the most infamous abuse and detraction, from personal insults, nor sometimes from real danger, arising from the insolent outrage of furious and disappointed monopolists.

On Foreign Trade

DAVID RICARDO

No extension of foreign trade will immediately increase the amount of value in a country, although it will very powerfully contribute to increase the mass of commodities, and therefore the sum of enjoyments. As the value of all foreign goods is measured by the quantity of the produce of our land and labour, which is given in exchange for them, we should have no greater value, if by the discovery of new markets, we obtained double the quantity of foreign goods in exchange for a given quantity of ours. . . .

It has indeed been contended, that the great profits which are sometimes made by particular merchants in foreign trade, will elevate the general rate of profits in the country, and that the abstraction of capital from other employments, to partake of the new and beneficial foreign commerce, will raise prices generally, and thereby increase profits. It has been said, by high authority, that less capital being necessarily devoted to the growth of corn, to the manufacture of cloth, hats, shoes, & c., while the demand continues the same, the price of these commodities will be so increased, that the farmer, hatter, clothier, and shoemaker, will have an increase of profits, as well as the foreign merchant.

They who hold this argument agree with me, that the profits of different employments have a tendency to conform to one another; to advance and recede together. Our variance consists in this: They contend that the equality of profits will be brought about by the general rise of profits; and I am of opinion, that the profits of the favoured trade will speedily subside to the general level.

Reprinted from David Ricardo, "On Foreign Trade," *The Principles of Political Economy and Taxation* (New York: E. P. Dutton & Co. Inc., 1960).

For, first, I deny that less capital will necessarily be devoted to the growth of corn, to the manufacture of cloth, hats, shoes, & c., unless the demand for these commodities be diminished; and if so, their price will not rise. In the purchase of foreign commodities, either the same, a larger, or a less portion of the produce of the land and labour of England will be employed. If the same portion be so employed, then will the same demand exist for cloth, shoes, corn, and hats as before, and the same portion of capital will be devoted to their production. If, in consequence of the price of foreign commodities being cheaper, a less portion of the annual produce of the land and labour of England is employed in the purchase of foreign commodities, more will remain for the purchase of other things. If their be a greater demand for hats, shoes, corn, & c., than before, which there may be, the consumers of foreign commodities having an additional portion of their revenue disposable, the capital is also disposable with which the greater value of foreign commodities was before purchased; so that with the increased demand for corn, shoes, & c., there exists also the means of procuring an increased supply, and therefore neither prices nor profits can permanently rise. If more of the produce of the land and labour of England be employed in the purchase of foreign commodities, less can be employed in the purchase of other things, and therefore fewer hats, shoes, & c., will be required. At the same time that capital is liberated from the production of shoes, hats, & c., more must be employed in manufacturing those commodities with which foreign commodities are purchased; and, consequently, in all cases the demand for foreign and home commodities together, as far as regards value, is limited by the revenue and capital of the country. If one increases the other must diminish. If the quantity of wine, imported in exchange for the same quantity of English commodities, be doubled, the people of England can either consume double the quantity of wine that they did before, or the same quantity of wine and a greater quantity of English commodities. If my revenue had been 1000£, with which I purchased annually one pipe of wine for 100£, and a certain quantiy of English commodities for 900£; when wine fell to 50£ per pipe, I might buy out the 50£ saved, either in the purchase of an additional pipe of wine, or in the purchase of more English commodities. If I bought more wine, and every wine-drinker did the same, the foreign trade would not be in the least disturbed; the same quantity of English commodities would be exported in exchange for wine, and we should receive double the quantity, though not double the value of wine. But if I, and others, contented ourselves with the same quantity of wine as before, fewer English commodities would be exported, and the wine-drinkers might either consume the commodities which were before exported, or any others for which they had an inclination. The capital required for their production would be supplied by the capital liberated from the foreign trade.

It is not, therefore, in consequence of the extension of the market that the rate of profit is raised, although such extension may be equally efficacious in increasing the mass of commodities, and may thereby enable us to augment the funds destined for the maintenance of labour, and the materials on which labour may be employed. It is quite as important to the happiness of mankind, that our enjoyments should be increased by the better distribution of labour, by each country producing those commodities for which by its situation, its climate, and its other natural or artificial advantages, it is adapted, and by their exchanging them for the commodities of other countries, as that they should be augmented by a rise in the rate of profits.

It has been my endeavour to show throughout this work, that the rate of profits can never be increased but by a fall in wages, and that there can be no permanent fall of wages but in consequence of a fall of the necessaries on which wages are expended. If, therefore, by the extension of foreign trade, or by improvements in machinery, the food and necessaries of the labourer can be brought to market, at a reduced price, profits will rise. If, instead of growing our own corn, or manufacturing the clothing and other necessaries of the labourer, we discover a new market from which we can supply ourselves with these commodities at a cheaper price, wages will fall and profits rise; but if the commodities obtained at a cheaper rate, by the extension of foreign commerce, or by the improvement of machinery, be exclusively the commodities consumed by the rich, no alteration will take place in the rate of profits. The rate of wages would not be affected, although wine, velvets, silks, and other expensive commodities should fall 50 per cent., and consequently profits would continue unaltered.

Foreign trade, then, though highly beneficial to a country, as it increases the amount and variety of the objects on which revenue may be expended, and affords, by the abundance and cheapness of commodities, incentives to saving, and to the accumulation of capital, has no tendency to raise the profits of stock, unless the commodities imported be of that description on which the wages of labour are expended.

The remarks which have been made respecting foreign trade, apply equally to home trade. The rate of profits is never increased by a better distribution of labour, by the invention of machinery, by the establishment of roads and canals, or by any means of abridging labour either in the manufacture or in the conveyance of goods. These are causes which operate on price, and never fail to be highly beneficial to consumers; since they enable them, with the same labour, or with the value of the produce of the same labour, to obtain in exchange a greater quantity of the commodity to which the improvement is applied; but they have no effect whatever on profit. On the other hand, every diminution in the wages of labour raises profits, but produces no effect on the price of commodities. One is advantageous to all classes, for all classes are consumers; the other is beneficial only to producers; they gain more, but every thing remains at its former price. In the first case they get the same as before; but every thing on which their gains are expended, is diminished in exchangeable value.

The same rule which regulates the relative value of commodities in one country, does not regulate the relative value of the commodities exchanged between two or more countries.

Under a system of perfectly free commerce, each country naturally devotes its capital and labour to such employments as are most beneficial to each. This pursuit of individual advantage is admirably connected with the universal good of the whole. By stimulating industry, by rewarding ingenuity, and by using most efficaciously the peculiar powers bestowed by nature, it distributes labour most effectively and most economically: while, by increasing the general mass of productions, it diffuses general benefit, and binds together, by one common tie of interest and intercourse, the universal society of nations throughout the civilized world. It is this principle which determines that wine shall be made in France and Portugal, that corn shall be grown in America and Poland, and that hardware and other goods shall be manufactured in England.

In one and the same country, profits are, generally speaking, always on the same level; or differ only as the employment of capital may be more or less secure and agreeable. It is not so between different countries. If the profits of capital employed in Yorkshire, should exceed those of capital employed in London, capital would speedily move from London to Yorkshire, and an equality of profits would be effected; but if in consequence of the diminished rate of production in the lands of England, from the increase of capital and population, wages should rise, and profits fall, it would not follow that capital and population would necessarily move from England to Holland, or Spain, or Russia, where profits might be higher.

If Portugal had no commercial connexion with other countries, instead of employing a great part of her capital and industry in the production of wines, with which she purchases for her own use the cloth and hardware of other countries, she would be obliged to devote a part of that capital to the manufacture of those commodities, which she would thus obtain probably inferior in quality as well as quantity.

The quantity of wine which she shall give in exchange for the cloth of England, is not determined by the respective quantities of labour devoted to the production of each, as it would be, if both commodities were manufactured in England, or both in Portugal.

England may be so circumstanced, that to produce the cloth may require the labour of 100 men for one year; and if she attempted to make the wine, it might require the labour of 120 men for the same time. England would therefore find it her interest to import wine, and to purchase it by the exportation of cloth.

To produce the wine in Portugal, might require only the labour of 80 men for one year, and to produce the cloth in the same country, might require the labour of 90 men for the same time. It would therefore be advantageous for her to export wine in exchange for cloth. This exchange might even take place, notwithstanding that the commodity imported by Portugal could be produced there with less labour than in England. Though she could make the cloth with the labour of 90 men, she would import it from a country where it required the labour of 100 men to produce it, because it would be advantageous to her rather to employ her capital in the production of wine, for which she would obtain more cloth from England, than she could produce by diverting a portion of her capital from the cultivation of vines to the manufacture of cloth.

Thus England would give the produce of the labour of 100 men, for the produce of the labour of 80. Such an exchange could not take place between the individuals of the same country. The labour of 100 Englishmen cannot be given for that of 80 Englishmen, but the produce of the labour of 100 Englishmen may be given for the produce of the labour of 80 Portuguese, 60 Russians, or 120 East Indians. The difference in this respect, between a single country and many, is easily accounted for, by considering the difficulty with which capital moves from one country to another, to seek a more profitable employment, and the activity with which it invariably passes from one province to another in the same country.

It would undoubtedly be advantageous to the capitalists of England, and to the consumers in both countries, that under such circumstances, the wine and the cloth should both be made in Portugal, and therefore that the capital and labour of England employed in making cloth, should be removed to Portugal for that purpose. In that case, the relative value of these commodities would be regulated by the same principle, as if one were the produce of Yorkshire, and the other of London: and in every other case, if capital freely flowed towards those countries where it could be most profitably employed, there could be no difference in the rate of profit, and no other difference in the real or labour price of commodities, than the additional quantity of labour required to convey them to the various markets where they were to be sold.

Experience, however, shows, that the fancied or real insecurity of capital, when not under the immediate control of its owner, together with the natural disinclination which every man has to quit the country of his birth and connexions, and intrust himself, with all his habits fixed, to a strange government and new laws, check the emigration of capital. These feelings, which I should be sorry to see weakened, induce most men of property to be satisfied with a low rate of profits in their own country, rather than seek a more advantageous employment for their wealth in foreign nations.

Gold and silver having been chosen for the general medium of circulation, they are, by the competition of commerce, distributed in such proportions amongst the different countries of the world, as to accommodate themselves to the natural traffic which would take place if no such metals existed, and the trade between countries were purely a trade of barter.

Thus, cloth cannot be imported into Portugal, unless it sell there for more gold than it cost in the country from which it was imported; and wine cannot be imported into England, unless it will sell for more there than it cost in Portugal. If the trade were purely a trade of barter, it could only continue whilst England could make cloth so cheap as to obtain a greater quantity of wine with a given quantity of labour, by manufacturing cloth than by growing vines; and also whilst the industry of Portugal were attended by the reverse effects. Now suppose England to discover a process for making wine, so that it should become her interest rather to grow it than import it; she would naturally divert a portion of her capital from the foreign trade to the home trade; she would cease to manufacture cloth for exportation, and would grow wine for herself. The money price of these commodities would be regulated accordingly; wine would fall here while cloth continued at its former price, and in Portugal no alteration would take place in the price of either commodity. Cloth would continue for some time to be exported from this country, because its price would continue to be higher in Portugal than here; but money instead of wine would be given in exchange for it, till the accumulation of money here, and its diminution abroad, should so operate on the relative value of cloth in the two countries, that it would cease to be profitable to export it. If the improvement in making wine were of a very important description, it might become profitable for the two countries to exchange employments; for England to make all the wine, and Portugal all the cloth consumed by them; but this could be effected only by a new distribution of the precious metals, which should raise the price of cloth in England, and lower it in Portugal. The relative price of wine would fall in England in consequence of the real advantage from the improvement of its manufacture; that is to say, its natural price would fall; the relative price of cloth would rise there from the accumulation of money.

But the diminution of money in one country, and its increase in another, do not operate on the price of one commodity only, but on the prices of all, and therefore the price of wine and cloth will be both raised in England, and both lowered in Portugal. . . .

It is thus that the money of each country is apportioned to it in such quantities only as may be necessary to regulate a profitable trade of barter. England exported cloth in exchange for wine, because, by so doing, her industry was rendered more productive to her; she had more cloth and wine than if she had manufactured both for herself; and Portugal imported cloth and exported wine, because the industry of Portugal could be more beneficially employed for both countries in producing wine. Let there be more difficulty in England in producing cloth, or in Portugal in producing wine, or let there be more facility in England in producing wine, or in Portugal in producing cloth, and the trade must immediately cease.

No change whatever takes place in the circumstances of Portugal; but England finds that she can employ her labour more productively in the manufacture of wine, and instantly the trade of barter between the two countries changes. Not only is the exportation of wine from Portugal stopped, but a new distribution of the precious metals takes place, and her importation of cloth is also prevented.

Both countries would probably find it their interest to make their own wine and their own cloth; but this singular result would take place: in England, though wine would be cheaper, cloth would be elevated in price, more would be paid for it by the consumer; while in Portugal the consumers, both of cloth and of wine, would be able to purchase those commodities cheaper. In the country where the improvement was made, prices would be enhanced; in that where no change had taken place, but where they had been deprived of a profitable branch of foreign trade, prices would fall.

This, however, is only a seeming advantage to Portugal, for the quantity of cloth and wine together produced in that country would be diminished, while the quantity produced in England would be increased. Money would in some degree have changed its value in the two countries; it would be lowered in England and raised in Portugal. Estimated in money, the whole revenue of Portugal would be diminished; estimated in the same medium, the whole revenue of England would be increased.

Thus, then, it appears that the improvement of a manufacture in any country tends to alter the distribution of the precious metals amongst the nations of the world: it tends to increase the quantity of commodities, at the same time that it raises general prices in the country where the improvement takes place.

To simplify the question, I have been supposing the trade between two countries to be confined to two commodities—to wine and cloth; but it is well known that many and various articles enter into the list of exports and imports. By the abstraction of money from one country, and the accumulation of it in another, all commodities are affected in price, and consequently encouragement is given to the exportation of many more commodities besides money, which will therefore prevent so great an effect from taking place on the value of money in the two countries as might otherwise be expected.

Beside the improvements in arts and machinery, there are various other causes which are constantly operating on the natural course of trade, and which interfere with the equilibrium, and the relative value of money. Bounties on exportation or importation, new taxes on commodities, sometimes by their direct, and at other times by their indirect operation, disturb the natural trade of barter, and produce a consequent necessity of importing or exporting money, in order that prices may be accommodated to the natural course of commerce; and this effect is produced not only in the country where the disturbing cause takes place, but, in a greater or less degree, in every country of the commercial world.

This will in some measure account for the different value of money in different countries; it will explain to us why the prices of home commodities, and those of great bulk, though of comparatively small value, are, independently of other causes, higher in those countries where manufactures flourish. Of two countries having precisely the same population, and the same quantity of land of equal fertility in cultivation, with the same knowledge too of agriculture, the prices of raw produce will be highest in that where the greater skill, and the better machinery is used in the manufacture of exportable commodities. The rate of profits will probably differ but little; for wages, or the real reward of the labourer, may be the same in both; but those wages, as well as raw produce, will be rated higher in money in that country, into which, from the advantages attending their skill and machinery, an abundance of money is imported in exchange for their goods.

Of these two countries, if one had the advantage in the manufacture of goods of one quality, and the other in the manufacture of goods of another quality, there would be no decided influx of the precious metals into either; but if the advantage very heavily preponderated in favour of either, that effect would be inevitable.

In the former part of this work, we have assumed, for the purpose of argument, that money always continued of the same value; we are now endeavouring to show that, besides the ordinary variations in the value of money, and those which are common to the whole commercial world, there are also partial variations to which money is subject in particular countries; and to the fact, that the value of money is never the same in any two countries, depending as it does on relative taxation, on manufacturing skill, on the advantages of climate, natural productions, and many other causes.

Although, however, money is subject to such perpetual variations, and consequently the prices of the commodities which are common to most countries, are also subject to considerable difference, yet no effect will be produced on the rate of profits, either from the influx or efflux of money. Capital will not be increased, because the circulating medium is agumented. If the rent paid by the farmer to his landlord, and the wages to his labourers, be 20 per cent. higher in one country than another, and if at the same time the nominal value of the farmer's capital be 20 per cent. more, he will receive precisely the same rate of profits, although he should sell his raw produce 20 per cent. higher.

Profits, it cannot be too often repeated, depend on wages; not on nominal, but real wages; not on the number of pounds that may be annually paid to the labourer, but on the number of days' work necessary to obtain those pounds. Wages may therefore be precisely the same in two countries; they may bear, too, the same proportion to rent, and to the whole produce obtained from the land, although in one of those countries the labourer should receive ten shillings per week, and in the other twelve.

In the early states of society, when manufactures have made little progress, and the produce of all countries is nearly similar, consisting of the bulky and most useful commodities, the value of money in different countries will be chiefly regulated by their distance from the mines which supply the precious metals; but as the arts and improvements of society advance, and different nations excel in particular manufactures, although distance will still enter into the calculation, the value of the precious metals will be chiefly regulated by the superiority of those manufactures.

Suppose all nations to produce corn, cattle, and coarse clothing only, and that it was by the exportation of such commodities that gold could be obtained from the countries which produced them, or from those who held them in subjection; gold would naturally be of greater exchangeable value in Poland than in England, on account of the greater expense of sending such a bulky commodity as corn the more distant voyage, and also the greater expense attending the conveying of gold to Poland.

This difference in the value of gold, or, which is the same thing, this difference in the price of corn in the two countries, would exist, although the facilities of producing corn in England should far exceed those of Poland, from the greater fertility of the land, and the superiority in the skill and implements of the labourer.

If, however, Poland should be the first to improve her manufactures, if she

should succeed in making a commodity which was generally desirable, including great value in little bulk, or if she should be exclusively blessed with some
natural production, generally desirable, and not possessed by other countries,
she would obtain an additional quantity of gold in exchange for this commodity, which would operate on the price of her corn, cattle, and coarse clothing.
The disadvantage of distance would probably be more than compensated by
the advantage of having an exportable commodity of great value, and money
would be permanently of lower value in Poland than in England. If, on the
contrary, the advantage of skill and machinery were possessed by England,
another reason would be added to that which before existed, why gold should
be less valuable in England than in Poland, and why corn, cattle, and clothing,
should be at a higher price in the former country.

These I believe to be the only two causes which regulate the comparative
value of money in the different countries of the world; for although taxation
occasions a disturbance of the equilibrium of money, it does so by depriving
the country in which it is imposed of some of the advantages attending skill,
industry and climate.

It has been my endeavour carefully to distinguish between a low value of
money, and a high value of corn, or any other commodity with which money
may be compared. These have been generally considered as meaning the
same thing; but it is evident, that when corn rises from five to ten shillings a
bushel, it may be owing either to a fall in the value of money, or to a rise in the
value of corn. Thus we have seen, that from the necessity of having recourse
successively to land of a worse and worse quality, in order to feed an increasing population, corn must rise in relative value to other things. If therefore
money continue permanently of the same value, corn will exchange for more
of such money, that is to say, it will rise in price. The same rise in the price of
corn will be produced by such improvement of machinery in manufactures, as
shall enable us to manufacture commodities with peculiar advantages: for the
influx of money will be the consequence; it will fall in value, and therefore
exchange for less corn. But the effects resulting from a high price of corn
when produced by the rise in the value of corn, and when caused by a fall in
the value of money, are totally different. In both cases the money price of
wages will rise, but if it be in consequence of the fall in the value of money, not
only wages and corn, but all other commodities will rise. If the manufacturer
has more to pay for wages, he will receive more for his manufactured goods,
and the rate of profits will remain unaffected. But when the rise in the price of
corn is the effect of the difficulty of production, profits will fall; for the
manufacturer will be obliged to pay more wages, and will not be enable to
remunerate himself by raising the price of his manufactured commodity.

Any improvement in the facility of working the mines, by which the precious metals may be produced with a less quantity of labour, will sink the value
of money generally. It will then exchange for fewer commodities in all countries; but when any particular country excels in manufactures, so as to occasion
an influx of money towards it, the value of money will be lower, and the prices
of corn and labour will be relatively higher in that country than in any other.

This higher value of money will not be indicated by the exchange; bills may continue to be negotiated at par, although the prices of corn and labour should be 10, 20, or 30 percent. higher in one country than another. Under the circumstances supposed, such a difference of prices is the natural order of things, and the exchange can only be at par, when a sufficient quantity of money is introduced into the country excelling in manufactures, so as to raise the price of its corn and labour. If foreign countries should prohibit the exportation of money, and could successfully enforce obedience to such a law, they might indeed prevent the rise in the prices of the corn and labour of the manufacturing country; for such rise can only take place after the influx of the precious metals, supposing paper money not to be used; but they could not prevent the exchange from being very unfavourable to them. If England were the manufacturing country, and it were possible to prevent the importation of money, the exchange with France, Holland, and Spain, might be 5, 10, or 20 per cent. against those countries.

Whenever the current of money is forcibly stopped, and when money is prevented from settling at its just level, there are no limits to the possible variations of the exchange. The effects are similar to those which follow, when a paper money, not exchangeable for specie at the will of the holder, is forced into circulation. Such a currency is necessarily confined to the country where it is issued: it cannot, when too abundant, diffuse itself generally amongst other countries. The level of circulation is destroyed, and the exchange will inevitably be unfavourable to the country where it is excessive in quantity: just so would be the effects of a metallic circulation, if by forcible means, by laws which could not be evaded, money should be detained in a country, when the stream of trade gave it an impetus towards other countries.

When each country has precisely the quantity of money which it ought to have, money will not indeed be of the same value in each, for with respect to many commodities it may differ 5, 10, or even 20 per cent., but the exchange will be at par. One hundred pounds in England, or the silver which is in 100£, will purchase a bill of 100£, or an equal quantity of silver in France, Spain, or Holland.

In speaking of the exchange and the comparative value of money in different countries, we must not in the least refer to the value of money estimated in commodities, in either country. The exchange is never ascertained by estimating the comparative value of money in corn, cloth, or any commodity whatever, but by estimating the value of the currency of one country, in the currency of another.

. . . How is it to be ascertained whether English money has fallen, or Hamburg money has risen? There is no standard by which this can be determined. It is a plea which admits of no proof, and can neither be positively affirmed, nor positively contradicted. The nations of the world must have been early convinced, that there was no standard of value in nature, to which they might unerringly refer, and therefore chose a medium, which on the whole appeared to them less variable than any other commodity.

To this standard we must conform till the law is changed, and till some other commodity is discovered, by the use of which we shall obtain a more perfect standard than that which we have established. While gold is exclusively the standard in this country, money will be depreciated, when a pound sterling is not of equal value with 5 dwts. and 3 grs. of standard gold, and that, whether gold rises or falls in general value.

3

Marx and the Early Marxists

Karl Marx (1818–1883) was not primarily interested in international political economy. His greatest work examined the genesis and internal dynamics of the capitalist mode of production. He commented upon the expansion of capital internationally, especially British colonialism, but his thoughts on the subject were not rigorously integrated into his general theory of capital. Marx's episodic writing on IPE centers on two themes: the dialectical interaction of international forces and the development of capitalism; and the consequences of the internationalization of capital.

Marx has been interpreted as holding an underconsumptionist theory of imperialism. An often-quoted passage from the *Communist Manifesto* appears to bear this out: "The need of a constantly expanding market for its products chases the bourgeoisie over the whole surface of the globe." This has been taken to mean that as capitalism develops, and the working class becomes more and more impoverished, domestic demand decreases. In such circumstances, the accumulation of capital, which is dependent upon the realization of surplus value through the sale of commodities, is undermined. Thus, if accumulation is to continue, and the capitalist mode of production is to survive, capitalists must seek overseas markets to sell their goods. Underconsumption at home forces capital to expand abroad. This argument implies that expansion is necessary for a relatively well-developed capitalist mode of production, one in which class contradictions between bourgeoisie and proletariat have matured to the point of underconsumption.

In other writings, however, Marx makes quite a different argument. His analysis of the "Rise of Manufactures" in the *German Ideology* illustrates how capital is not pushed overseas by internal contradictions; it is pulled by new opportunities.

Marx posits three phases in the development of manufacturing industry. The first period, running from the advent of a manufacturing division of labor in roughly the fifteenth century to the early sixteenth century, is transitionary in nature. Feudal class structures and economic practices are extant, but some manufacturers are beginning to break away from the medieval guild system. This process of socioeconomic transformation is promoted by, and itself promotes, international trade. In creating new sources of wealth, foreign trade promotes the rise of the bourgeoisie. New domestic conditions, in turn, transform international economic relations. Where once trade was "inoffensive exchange," it is now highly competitive, marked by mercantilism and colonization. This competitiveness cannot be explained by underconsumption; the capitalist mode of production is hardly established. Interestingly, Marx suggests that early mercantilism is due to state interests: "The state, which was daily less and less able to do without money, now retained the ban on the export of gold and silver out of fiscal considerations." Although this ultimately serves the interests of capital, as manufacturing grows behind protectionist barriers, it opens the possibility of relatively autonomous state action.

The second phase in the development of manufacturing is mature mercantilism and the rise of the "big bouregoisie," extending from the mid-seventeenth to the late eighteenth century. The merchant class, taking greatest advantage of protectionism, forms the most powerful sector of the bourgeoisie, while manufacturers are small-scale and possessed of "an extreme petty-bourgeois outlook." The concentration of trade and manufacturing in Britain at this time, however, is of pivotal importance. British interests create a world market for their goods, building "a demand for the manufactured products of this country, which could no longer be met by the industrial productive forces hitherto existing." In short, demand is greater than supply, and this stimulates the further development of domestic industry and international trade. Again, industrial capital has yet to fully mature; therefore, the underconsumptionist argument does not adequately explain these phenomena.

The expansion of international demand ushers in the third phase of manufacturing development. This period, covering the nineteenth century, is charcterized by universalized international competition and the ascendance of manufacturers over merchants. Protectionism becomes merely a rearguard action against the onslaught of free trade. During this time, the full maturation of industrial capital occurs and, presumably, underconsumption comes into play, though Marx does not argue this explicitly. In sum, the relationship of the emergence of capital and foreign economic relations is dialectical and interactive.

Marx also analyzes the consequences of the internationalization of capital. In both the *Communist Manifesto* and the *German Ideology,* he suggests that capitalism will ultimately expand and dominate the world: "The cheap prices of its commodities are the heavy artillery with which it batters down all Chinese walls." His discussion of India provides greater detail to this point.

Domestic economic, social, and political conditions in India are very different from the European experience. The indigenous "village system," which Marx elsewhere refers to as the Asiatic mode of production, is composed of a powerful centralized state perched above, and largely autonomous from, an atomized collection of self-reliant agricultural towns. The lack of organic connections among the towns, and between state and society, rob this peculiar amalgam of the developmental dynamism of European modes of production. Asia is simply resistant to social change from within. But British colonialism destroys the structural foundation of the Asiatic mode of production. India will industrialize and class relations will be restructured. Even "changeless" Asia cannot resist the transforming power of capital. Although this is a brutal process, selfishly serving the interests of the British bourgeoisie, it is inevitable and necessary to create conditions for further development in India.

Vladimir Lenin (1870–1924) agrees with Marx's conclusion that capital will expand and dominate the world. However, Lenin's discussion of the causes of capitalist expansion differs somewhat from Marx's broad historical analysis. Writing in the twentieth century, Lenin stresses the uneven nature of capitalist development that leads him to an underconsumptionist position: "The necessity of exporting capital arises from the fact that in a few countries capitalism has become "overripe" and (owing to the backward state of agriculture and the impoverished state of the masses) capital cannot field 'profitable' investment." From this premise, Lenin argues that imperialism is a distinct stage, indeed the "highest stage," of capitalism. Building upon Rudolf Hilferding's concept of "finance capital," Lenin illustrates how concentration and centralization eventually require the internationalization of capital. This has crucial political consequences, as the imperialist powers first "divide the world" among themselves, and are finally driven to war among themselves. Although Lenin's discussion of imperialism is derived from Marx's general analysis of capital, and has greatly influenced succeeding generations of Marxists, its underconsumptionist inclination does not fully capture the complexity of Marx's examination of the ways in which international economic forces shaped the earlier rise of capitalism.

Manifesto Of The Communist Party

KARL MARX AND FRIEDRICH ENGELS

I. Bourgeois and Proletarians

The history of all hitherto existing society is the history of class struggles.

The modern bourgeois society that has sprouted from the ruins of feudal society, has not done away with class antagonisms. It has but established new classes, new conditions of oppression, new forms of struggle in place of the old ones.

Our epoch, the epoch of the bourgeoisie, possesses, however, this distinctive feature: It has simplified the class antagonisms. Society as a whole is more and more splitting up into two great hostile camps, into two great classes directly facing each other—bourgeoisie and proletariat.

From the serfs of the Middle Ages sprang the chartered burghers of the earliest towns. From these burgesses the first eleménts of the bourgeoisie were developed.

The discovery of America, the rounding of the Cape, opened up fresh ground for the rising bourgeoisie. The East Indian and Chinese markets, the colonization of America, trade with the colonies, the increase in the means of exchange and in commodities generally, gave to commerce, to navigation, to industry, an impulse never before known, and thereby, to the revolutionary element in the tottering feudal society, a rapid development.

The feudal system of industry, in which industrial production was monopolized by closed guilds, now no longer sufficed for the growing wants of the new markets. The manufacturing system took its place. The guild-masters were pushed aside by the manufacturing middle class; division of labour between the different corporate guilds vanished in the face of division of labour in each single workshop.

Meantime the markets kept ever growing, the demand ever rising. Even manufacture no longer sufficed. Thereupon, steam and machinery revolutionized industrial production. The place of manufacture was taken by the giant, modern industry, the place of the industrial middle class, by industrial millionaires—the leaders of whole industrial armies, the modern bourgeois.

Modern industry has established the world market, for which the discovery of America paved the way. This market has given an immense development to commerce, to navigation, to communication by land. This development has, in its turn, reacted on the extension of industry; and in proportion as industry, commerce, navigation, railways extended, in the same proportion the bourgeoisie developed, increased its capital, and pushed into the background every class handed down from the Middle Ages.

We see, therefore, how the modern bourgeoisie is itself the product of a long course of development, of a series of revolutions in the modes of production and of exchange.

Each step in the development of the bourgeoisie was accompanied by a corresponding political advance of that class. An oppressed class under the sway of the feudal nobility, it became an armed and self-governing association in the medieval commune; here independent urban republic (as in Italy and Germany), there taxable "third estate" of the monarchy (as in France); afterwards, in the period of manufacture proper, serving either the semi-feudal or the absolute monarchy as a counterpoise against the nobility, and, in fact, cornerstone of the great monarchies in general—the bourgeoisie has at last, since the establishment of modern industry and of the world market, conquered for itself, in the modern representative state, exclusive political sway. The executive of the modern state is but a committee for managing the common affairs of the whole bourgeoisie.

The bourgeoisie has played a most revolutionary role in history.

The bourgeoisie, wherever it has got the upper hand, has put an end to all feudal, patriarchal, idyllic relations. It has pitilessly torn asunder the motley feudal ties that bound man to his "natural superiors," and has left no other bond between man and man than naked self-interest, than callous "cash payment." It has drowned the most heavenly ecstasies of religious fervor, of chivalrous enthusiasm, of philistine sentimentalism, in the icy water of egotistical calculation. It has resolved personal worth into exchange value, and in place of the numberless indefeasible chartered freedoms, has set up that single, unconscionable freedom—Free Trade. In one word, for exploitation, veiled by religious and political illusions, it has substituted naked, shameless, direct, brutal exploitation.

The bourgeoisie cannot exist without constantly revolutionizing the instruments of production, and thereby the relations of production, and with them the whole relations of society. Conservation of the old modes of production in unaltered form, was, on the contrary, the first condition of existence for all earlier industrial classes. Constant revolutionizing of production, uninterrupted disturbance of all social conditions, everlasting uncertainty and agitation distinguish the bourgeois epoch from all earlier ones. All fixed, fast-frozen relations, with their train of ancient and venerable prejudices and opinions, are swept away, all new-formed ones become antiquated before they can ossify. All that is solid melts into air, all that is holy is profaned, and man is at last compelled to face with sober senses his real conditions of life and his relations with his kind.

The need of a constantly expanding market for its products chases the bourgeoisie over the whole surface of the globe. It must nestle everywhere, settle everywhere, establish connections everywhere.

The bourgeoisie has through its exploitation of the world market given a cosmopolitan character to production and consumption in every country. To the great chagrin of reactionaries, it has drawn from under the feet of industry the national ground on which it stood. All old-established national industries have been destroyed or are daily being destroyed. They are dislodged by new industries, whose introduction becomes a life and death question for all civilized nations, by industries that no longer work up indigenous raw material, but raw material drawn from the remotest zones; industries whose products are consumed, not only at home, but in every quarter of the globe. In place of the old wants, satisfied by the production of the country, we find new wants, requiring for their satisfaction the products of distant lands and climes. In place of the old local and national seclusion and self-sufficiency, we have intercourse in every direction, universal inter-dependence of nations. And as in material, so also in intellectual production. The intellectual creations of individual nations become common property. National one-sidedness and narrow-mindedness become more and more impossible, and from the numerous national and local literatures there arises a world literature.

The bourgeoisie, by the rapid improvement of all instruments of production, by the immensely facilitated means of communication, draws all nations, even the most barbarian, into civilization. The cheap prices of its commodities are the heavy artillery with which it batters down all Chinese walls, with which it forces the barbarians' intensely obstinate hatred of foreigners to capitulate. It compels all nations, on pain of extinction, to adopt the bourgeois mode of production; it compels them to introduce what it calls civilization into their midst, i.e., to become bourgeois themselves. In a word, it creates a world after its own image.

The bourgeoisie has subjected the country to the rule of the towns. It has created enormous cities, has greatly increased the urban population as compared with the rural, and has thus rescued a considerable part of the population from the idiocy of rural life. Just as it has made the country dependent on the towns, so it has made barbarian and semi-barbarian countries dependent on the civilized ones, nations of peasants on nations of bourgeois, the East on the West.

More and more the bourgeoisie keeps doing away with the scattered state of the population, of the means of production, and of property. It has agglomerated population, centralized means of production, and has concentrated property in a few hands. The necessary consequence of this was political centralization. Independent, or but loosely connected provinces, with separate interests, laws, governments, and systems of taxation, became lumped together into one nation, with one government, one code of laws, one national class interest, one frontier, and one customs tariff.

The German Ideology:
The Rise of Manufactures

KARL MARX

With the advent of manufactures, the various nations entered into a competitive relationship, the struggle for trade, which was fought out in wars, protective duties and prohibitions, whereas earlier the nations, insofar as they were connected at all, had carried on an inoffensive exchange with each other. Trade had from now on a political significance.

Manufacture and the movement of production in general received an enormous impetus through the extension of commerce which came with the discovery of America and the sea-route to the East Indies. The new products imported thence, particularly the masses of gold and silver which came into circulation and totally changed the position of the classes towards one another, dealing a hard blow to feudal landed property and to workers; the expeditions of adventurers, colonisation; and above all the extension of markets into a world market, which has now become possible and was daily becoming more and more a fact, called forth a new phase of historical development, into which in general we cannot here enter further. Through the colonisation of the newly discovered countries the commercial struggle of the nations amongst one another was given new fuel and accordingly greater extension and animosity.

The expansion of trade and manufacture accelerated the accumulation of movable capital, while in the guilds, which were not stimulated to extend their production, natural capital remained stationary or even declined. Trade and manufacture created the big bourgeoisie; in the guilds was concentrated the petty bourgeoisie, which no longer was dominant in the towns as formerly, but

Karl Marx, *The German Ideology* (New York: International Publishers, 1947).

had to bow to the might of the great merchants and manufacturers. Hence the decline of the guilds, as soon as they came into contact with manufacture.

The intercourse of nations took on, in the epoch of which we have been speaking, two different forms. At first the small quantity of gold and silver in circulation involved the ban on the export of these metals; and industry, for the most part imported from abroad and made necessary by the need for employing the growing urban population, could not do without those privileges which could be granted not only, of course, against home competition, but chiefly against foreign. The local guild privilege was in these original prohibitions extended over the whole nation. Customs duties originated from the tributes which the feudal lords exacted as protective levies against robbery from merchants passing through their territories, tributes later imposed likewise by the towns, and which, with the rise of the modern states, were the Treasury's most obvious means of raising money.

The appearance of American gold and silver on the European markets, the gradual development of industry, the rapid expansion of trade and the consequent rise of the non-guild bourgeoisie and of money, gave these measures another significance. The State, which was daily less and less able to do without money, now retained the ban on the export of gold and silver out of fiscal considerations; the bourgeois, for whom these masses of money which were hurled on to the market became the chief object of speculative buying, were thoroughly content with this; privileges established earlier became a source of income for the government and were sold for money; in the customs legislation there appeared the export duty, which, since it only [placed] a hindrance in the way of industry, had a purely fiscal aim.

The second period began in the middle of the seventeenth century and lasted almost to the end of the eighteenth. Commerce and navigation had expanded more rapidly than manufacture, which played a secondary role; the colonies were becoming considerable consumers; and after long struggles the separate nations shared out the opening world market among themselves. This period begins with the Navigation Laws and colonial monopolies. The competition of the nations among themselves was excluded as far as possible by tariffs, prohibitions and treaties; and in the last resort the competitive struggle was carried on and decided by wars (especially naval wars). The mightiest maritime nation, the English, retained preponderance in trade and manufacture. Here, already, we find concentration in one country.

Manufacture was all the time sheltered by protective duties in the home market, by monopolies in the colonial market, and abroad as much as possible by differential duties. The working-up of home-produced material was encouraged (wool and linen in England, silk in France), the export of home-produced raw material forbidden (wool in England), and the [working-up] of imported material neglected or suppressed (cotton in England). The nation dominant in sea trade and colonial power naturally secured for itself also the greatest quantitative and qualitative expansion of manufacture. Manufacture could not be carried on without protection, since, if the slightest change takes place in other countries, it can lose its market and be ruined; under reasonably favourable conditions it may easily be introduced into a country, but for this very reason can easily be destroyed. At the same time through the mode in which it is carried on, particularly in the eighteenth century, in the country-side, it is to such an extent interwoven with the vital relationships of a great mass of individuals, that no country dare jeopardise its existence by permitting free competition. Insofar as it manages to export, it therefore depends entirely on the extension or restriction of commerce, and exercises a relatively very small reaction [on the latter]. Hence its secondary [importance] and the influence of [the merchants] in the eighteenth century. It was the merchants and especially the shippers who more than anybody else pressed for State protection and monopolies; the manufacturers also demanded and indeed received protection, but all the time were inferior in political importance to the merchants. The commercial towns, particularly the maritime towns, became to some extent civilised and acquired the outlook of the big bourgeoisie, but in the factory towns an extreme petty-bourgeois outlook persisted.

This period is also characterised by the cessation of the bans on the export of gold and silver and the beginning of the trade in money; by banks, national debts, paper money, by speculation in stocks and shares and stockjobbing in all articles; by the development of finance in general. Again capital lost a great part of the natural character which had still clung to it.

The concentration of trade and manufacture in one country, England, developing irresistibly in the seventeenth century, gradually created for this country a relative world market, and thus a demand for the manufactured products of this country, which could no longer be met by the industrial productive forces hitherto existing. This demand, outgrowing the productive forces, was the motive power which, by producing big industry—the application of elemental forces to industrial ends, machinery and the most complex division of labour— called into existence the third period of private ownership since the Middle Ages. There already existed in England the other pre-conditions of this new phase: freedom of competition inside the nation, the development of theoretical mechanics, etc. (Indeed, the science of mechanics perfected by Newton was altogether the most popular science in France and England in the eighteenth century.) (Free competition inside the nation itself had everywhere to be conquered by a revolution—1640 and 1688 in England, 1789 in France.) Competition soon compelled every country that wished to retain its historical role to protect its manufactures by renewed customs regulations (the old duties were no longer any good against big industry) and soon after to introduce big industry under protective duties. Big industry universalised competition in spite of these protective measures (it is practical free trade; the protective duty is only a palliative, a measure of defence *within* free trade), established means of communication and the modern world market, subordinated trade to itself, transformed all capital into industrial capital, and thus produced the rapid circulation (development of the financial system) and the centralisation of capital. By universal competition it forced all individuals to strain their energy to the utmost. It destroyed as far as possible ideology, religion, morality, etc., and where it could not do this, made them into a palpable lie. It produced world history for the first time, insofar as it made all civilised nations and every individual member of them dependent for the satisfaction of their wants on the whole world, thus destroying the former natural exclusiveness of separate nations. It made natural science subservient to capital and took from the division of labour the last semblance of its natural character. It destroyed natural growth in general, as far as this is possible while labour exists, and resolved all natural relationships into money relationships. In the place of naturally grown towns it created the modern, large industrial cities which have sprung up overnight. Wherever it penetrated, it destroyed the crafts and all earlier stages of industry. It completed the victory of the commerical town over the countryside. Generally speaking, big industry created everywhere the same relations between the classes of society, and thus destroyed the peculiar individuality of the various nationalities. And finally, while the bourgeoisie of each nation still retained separate national interests, big industry created a class, which in all nations has the same interest and with which nationality is already dead; a class which is really rid of all the old world and at the same time stands pitted against it. Big industry makes for the worker not only the relation to the capitalist, but labour itself, unbearable.

It is evident that big industry does not reach the same level of development in all districts of a country. This does not, however, retard the class movement of the proletariat, because the proletarians created by big industry assume leadership of this movement and carry the whole mass along with them, and because the workers excluded from big industry are placed by it in a still worse situation than the workers in big industry itself. The countries in which big industry is developed act in a similar manner upon the more or less non-industrial countries, insofar as the latter are swept by universal commerce into the universal competitive struggle.

On Imperialism In India

KARL MARX

Hindostan is an Italy of Asiatic dimensions, the Himalayas for the Alps, the Plains of Bengal for the Plains of Lombardy, the Deccan for the Appenines, and the Isle of Ceylon for the Island of Sicily. The same rich variety in the products of the soil, and the same dismemberment in the political configuration. Just as Italy has, from time to time, been compressed by the conqueror's sword into different national masses, so do we find Hindostan, when not under the pressure of the Mohammedan, or the Mogul, or the Briton, dissolved into as many independent and conflicting States as it numbered towns, or even villages. Yet, in a social point of view, Hindostan is not the Italy, but the Ireland of the East. And this strange combination of Italy and of Ireland, of a world of voluptuousness and of a world of woes, is anticipated in the ancient traditions of the religion of Hindostan. That religion is at once a religion of sensualist exuberance, and a religion of self-torturing asceticism. . . .

There cannot, however, remain any doubt but that the misery inflicted by the British on Hindostan is of an essentially different and infinitely more intensive kind than all Hindostan had to suffer before.

All the civil wars, invasions, revolutions, conquests, famines, strangely complex, rapid and destructive as the successive action in Hindostan may appear, did not go deeper than its surface. England had broken down the entire framework of Indian society, without any symptoms of reconstitution yet appearing. This loss of his old world, with no gain of a new one, imparts a particular kind of melancholy to the present misery of the Hindoo, and sepa-

Reprinted from Robert C. Tucker, *The Marx-Engels Reader* (New York: Norton, 1978).

rates Hindostan, ruled by Britain, from all its ancient traditions, and from the whole of its past history.

There have been in Asia, generally, from immemorial times, but three departments of Government: that of Finance, or the plunder of the interior; that of War, or the plunder of the exterior; and, finally, the department of Public Works. Climate and territorial conditions, especially the vast tracts of desert, extending from the Sahara, through Arabia, Persia, India and Tartary, to the most elevated Asiatic highlands, constituted artificial irrigation by canals and waterworks the basis of Oriental agriculture. As in Egypt and India, inundations are used for fertilising the soil of Mesopotamia, Persia, etc.; advantage is taken of a high level for feeding irrigative canals. This prime necessity of an economical and common use of water, which, in the Occident, drove private enterprise to voluntary association, as in Flanders and Italy, necessitated, in the Orient where civilisation was too low and the territorial extent too vast to call into life voluntary association, the interference of the centralising power of Government. Hence an economical function devolved upon all Asiatic Governments the function of providing public works. This artificial fertilisation of the soil, dependent on a Central Government, and immediately decaying with the neglect of irrigation and drainage, explains the otherwise strange fact that we now find whole territories barren and desert that were once brilliantly cultivated, as Palmyra, Petra, the ruins in Yemen, and large provinces of Egypt, Persia and Hindostan; it also explains how a single war of devastation has been able to depopulate a country for centuries, and to strip it of all its civilisation.

Now, the British in East India accepted from their predecessors the department of finance and of war, but they have neglected entirely that of public works. Hence the deterioration of an agriculture which is not capable of being conducted on the British principle of free competition, of *laissez-faire* and *laissez-aller*. But in Asiatic empires we are quite accustomed to see agriculture deteriorating under one government and reviving again under some other government. There the harvests correspond to good or bad government, as they change in Europe with good or bad seasons. Thus the oppression and neglect of agriculture, bad as it is, could not be looked upon as the final blow dealt to Indian society by the British intruder, had it not been attended by a circumstance of quite different importance, a novelty in the annals of the whole Asiatic world. However changing the political aspect of India's past must appear, its social condition has remained unaltered since its remotest antiquity, until the first decennium of the 19th century. The hand-loom and the spinning-wheel, producing their regular myriads of spinners and weavers, were the pivots of the structure of that society. From immemorial times, Europe received the admirable textures of Indian labour, sending in return for them her precious metals, and furnishing thereby his material to the goldsmith, that indispensable member of Indian society, whose love of finery is so great that even the lowest class, those who go about nearly naked, have commonly a pair of golden ear-rings and a gold ornament of some kind hung round their necks. Rings on the fingers and toes have also been common.

Women as well as children frequently wore massive bracelets and anklets of gold or silver, and statuettes of divinities in gold and silver were met with in the households. It was the British intruder who broke up the Indian hand-loom and destroyed the spinning wheel. England began with driving the Indian cottons from the European market; it then introduced twist into Hindostan and in the end inundated the very mother country of cotton with cottons. From 1818 to 1836 the export of twist from Great Britain to India rose in the proportion of 1 to 5,200. In 1824 the export of British muslins to India hardly amounted to 1,000,000 yards while in 1837 surpassed 64,000,000 yards. But at the same time the population of Decca decreased from 150,000 inhabitants to 20,000. This decline of Indian towns celebrated for their fabrics was by no means the worst consequence. British steam and science uprooted, over the whole surface of Hindostan, the union between agricultural and manufacturing industry.

These two circumstances—the Hindoo, on the one hand, leaving, like all Oriental peoples, to the central government the care of the great public works, the prime condition of his agriculture and commerce, dispersed, on the other hand over the surface of the country, and agglomerated in small centers by the domestic union of agricultural and manufacturing pursuits—these two circumstances had brought about, since the remotest times, a social system of particular features—the so-called *village system,* which gave to each of these small unions their independent organisation and distinct life. . . .

These small stereotype forms of social organism have been to the greater part dissolved, and are disappearing, not so much through the brutal interference of the British tax-gatherer and the British soldier, as to the working of English steam and English free trade. Those family-communities were based on domestic industry, in that peculiar combination of hand-weaving, hand-spinning and hand-tilling agriculture which gave them self-supporting power. English interference having placed the spinner in Lancashire and the weaver in Bengal, or sweeping away both Hindoo spinner and weaver, dissolved these small semi-barbarian, semi-civilised communities, by blowing up their economical basis, and thus produced the greatest, and, to speak the truth, the only social revolution ever heard of in Asia.

Now, sickening as it must be to human feeling to witness those myriads of industrious patriarchal and inoffensive social organisations disorganised and dissolved into their units, thrown into a sea of woes, and their individual members losing at the same time their ancient form of civilisation, and their hereditary means of subsistence, we must not forget that these idyllic village communities, inoffensive though they may appear, had always been the solid foundation of Oriental despotism, that they restrained the human mind within the smallest possible compass, making it the unresisting tool of superstition, enslaving it beneath traditional rules, depriving it of all grandeur and historical energies. We must not forget the barbarian egotism which, concentrating on some miserable patch of land, had quietly witnessed the ruin of empires, the perpetration of unspeakable cruelties, the massacre of the population of large towns, with no other consideration bestowed upon them than on natural events, itself the helpless prey of any aggressor who deigned to notice it at all. We must not forget that this undignified, stagnatory, and vegetative life, that this passive sort of existence evoked on the other part, in contradistinction, wild, aimless, unbounded forces of destruction and rendered murder itself a religious rite in Hindostan. We must not forgeet that these little communities were contaminated by distinctions of caste and by slavery, that they subjugated man to external circumstances instead of elevating man to be the sovereign of circumstances, that they transformed a self-developing social state into never changing natural destiny, and thus brought about a brutalising worship of nature, exhibiting its degradation in the fact that man, the sovereign of nature, fell down on his knees in adoration of *Hunuman,* the monkey, and *Sabbala,* the cow.

England, it is true, in causing a social revolution in Hindostan, was actuated only by the vilest interests, and was stupid in her manner of enforcing them. But that is not the question. The question is, can mankind fulfill its destiny without a fundamental revolution in the social state of Asia? If not, whatever may have been the crimes of England she was the unconscious tool of history in bringing about that revolution.

The Export Of Capital

V. I. LENIN

Under the old capitalism, when free competition prevailed, the export of *goods* was the most typical feature. Under modern capitalism, when monopolies prevail, the export of *capital* has become the typical feature.

Capitalism is commodity production at the highest stage of development, when labour power itself becomes a commodity. The growth of internal exchange, and particularly of international exchange, is the characteristic distinguishing feature of capitalism. The uneven and spasmodic character of the development of individual enterprises, of individual branches of industry and individual countries, is inevitable under the capitalist system. England became a capitalist country before any other, and in the middle of the nineteenth century, having adopted free trade, claimed to be the "workshop of the world," the great purveyor of manufactured goods to all countries, which in exchange were to keep her supplied with raw materials. But in the last quarter of the nineteenth century, *this* monopoly was already undermined. Other countries, protecting themselves by tariff walls, had developed into independent capitalist states. On the threshold of the twentieth century, we see a new type of monopoly coming into existence. Firstly, there are monopolist capitalist combines in all advanced capitalist countries; secondly, a few rich countries, in which the accumulation of capital reaches gigantic proportions, occupy a monopolist position. An enormous "superabundance of capital" has accumulated in the advanced countries.

It goes without saying that if capitalism could develop agriculture, which

Reprinted from *Imperialism: The Highest State of Capitalism* by V. I. Lenin, "The Export of Capital," "Imperialism as a Special Stage of Capitalism," by permission of International Publishers, NY. (c) 1939.

today lags far behind industry everywhere, if it could raise the standard of living of the masses, who are everywhere still poverty-striken and underfed, in spite of the amazing advance in technical knowledge, there could be no talk of a superabundance of capital. This "argument" the petty-bourgeois critics of capitalism advance on every occasion. But if capitalism did these things it could not be capitalism; for uneven development and wretched conditions of the masses are fundamental and inevitable conditions and premises of this mode of production. As long as capitalism remains what it is, surplus capital will never be utilised for the purpose of raising the standard of living of the masses in a given country, for this would mean a decline in profits for the capitalist; it will be used for the purpose of increasing those profits by exporting capital abroad to the backward countries. In these backward countries profits are usually high, for capital is scarce, the price of land is relatively low, wages are low, raw materials are cheap. The possibility of exporting capital is created by the fact that numerous backward countries have been drawn into international capitalist intercourse; main railways have either been built or are being built there; the elementary conditions for industrial development have been created, etc. The necessity for exporting capital arises from the fact that in a few countries capitalism has become "over-ripe" and (owing to the backward state of agriculture and the impoverished state of the masses) capital cannot field "profitable" investment.

The export of capital greatly affects and accelerates the development of capitalism in those countries to which it is exported. While, therefore, the export of capital may tend to a certain extent to arrest development in the countries exporting capital, it can only do so by expanding and deepening the further development of capitalism throughout the world.

The countries which export capital are nearly always able to obtain "advantages," the character of which throws light on the peculiarities of the epoch of finance capital and monopoly. . . .

Finance capital has created the epoch of monopolies, and monopolies introduce everywhere monopolist methods: the utilisation of "connections" for profitable transactions takes the place of competition on the open market. The most usual thing is to stipulate that part of the loan that is granted shall be spent on purchases in the country of issue, particularly on orders for war materials, or for ships, etc. In the course of the last two decades (1890–1910), France often resorted to this method. The export of capital abroad thus becomes a means for encouraging the export of commodities. In these circumstances transactions between particularly big firms assume a form "bordering on corruption," as Schilder "delicately" puts it. Krupp in Germany, Schneider in France, Armstrong in England are instances of firms which have close connections with powerful banks and governments and cannot be "ignored" when arranging a loan.

Thus, finance capital, almost literally, one might say, spreads its net over all countries of the world. Banks founded in the colonies or their branches, play an important part in these operations. German imperialists look with envy on the "old" colonising nations which are "well established" in this respect. In 1904, Great Britain had 50 colonial banks with 2,279 branches (in 1910 there were 72 banks with 5,449 branches); France had 20 with 136 branches; Holland 16 with 68 branches; and Germany had a "mere" 13 with 70 branches. The American capitalists, in their turn, are jealous of the English and German: "In South America," they complained in 1915, "five German banks have forty branches and five English banks have seventy branches. . . . England and Germany have invested in Argentina, Brazil, and Uruguay in the last twenty-five years approximately four thousand million dollars, and as a result enjoy together 46 per cent of the total trade of these three countries."

The capital exporting countries have divided the world among themselves in the figurative sense of the term. But finance capital has also led to the *actual* division of the world.

Imperialism as a Special Stage of Capitalism

V. I. LENIN

We must now try to sum up and put together what has been said above on the subject of imperialism. Imperialism emerged as the development and direct continuation of the fundamental attributes of capitalism in general. But capitalism only became capitalist imperialism at a definite and very high stage of its development when certain of its fundamental attributes began to be transformed into their opposites, when the features of a period of transition from capitalism to a higher social and economic system began to take shape and reveal themselves all along the line. Economically, the main thing in this process is the substitution of capitalist monopolies for capitalist free competition. Free competition is the fundamental attribute of capitalism, and of commodity production generally. Monopoly is exactly the opposite of free competition; but we have seen the latter being transformed into monopoly before our very eyes, creating large-scale industry and eliminating small industry, replacing large-scale industry by still larger-scale industry, finally leading to such a concentration of production and capital that monopoly has been and is the result: cartels, syndicates and trusts, and merging with them, the capital of a dozen or so banks manipulating thousands of millions. At the same time monopoly, which has grown out of free competition, does not abolish the latter, but exists over it and alongside of it, and thereby gives rise to a number of very acute, intense antagonisms, frictions and conflicts. Monopoly is the transition from capitalism to a higher system.

If it were necessary to give the briefest possible definition of imperialism we should have to say that imperialism is the monopoly stage of capitalism. Such a definition would include what is most important, for, on the one hand, finance capital is the bank capital of a few big monopolist banks, merged with the capital of the monopolist combines of manufacturers; and, on the other hand, the division of the world is the transition from a colonial policy which has extended without hindrance to territories unoccupied by any capitalist power, to a colonial policy of monopolistic possession of the territory of the world which has been completely divided up.

But very brief definitions, although convenient, for they sum up the main points, are nevertheless inadequate, because very important features of the phenomenon that has to be defined have to be especially deduced. And so, without forgetting the conditional and relative value of all definitions, which can never include all the concatenations of a phenomenon in its complete development, we must give a definition of imperialism that will embrace the following five essential features:

1) The concentration of production and capital developed to such a high stage that it created monopolies which play a decisive role in economic life.

2) The merging of bank capital with industrial capital, and the creation, on the basis of this "finance capital," of a "financial oligarchy."

3) The export of capital, which has become extremely important, as distinguished from the export of commodities.

4) The formation of international capitalist monopolies which share the world among themselves.

5) The territorial division of the whole world among the greatest capitalist powers is completed.

Imperialism is capitalism in that stage of development in which the dominance of monopolies and finance capital has established itself; in which the export of capital has acquired pronounced importance; in which the division of the world among the international trusts has begun; in which the division of all territories of the globe among the great capitalist powers has been completed.

In this matter of defining imperialism, however, we have to enter into controversy, primarily, with K. Kautsky, the principal Marxian theoretician of the epoch of the so-called Second International—that is, of the twenty-five years between 1889 and 1914.

Kautsky, in 1915 and even in November 1914, very emphatically attacked the fundamental ideas expressed in our definition of imperialism. Kautsky said that imperialism must not be regarded as a "phase" or stage of economy, but as a policy; a definite policy "preferred" by finance capital; that imperialism cannot be "identified" with "contemporary capitalism"; that if imperialism is to be understood to mean "all the phenomena of contemporary capitalism"—cartels, protection, the domination of the financiers and colonial policy—then the question as to whether imperialism is necessary to capitalism becomes reduced to the "flattest tautology"; because, in that case, "imperialism is naturally a vital necessity for capitalism," and so on. The best way to present Kautsky's ideas is to quote his own definition of imperialism, which is diametrically opposed to the substance of the ideas which we have set forth. . . .

Kautsky's definition is as follows:

> Imperialism is a product of highly developed industrial capitalism. It consists in the striving of every industrial capitalist nation to bring under its control and to annex increasingly big *agrarian*" (Kautsky's italics) "regions irrespective of what nation inhabit those regions."

This definition is utterly worthless because it one-sidedly, i.e., arbitrarily, brings out the national question alone (although this is extremely important in itself as well as in its relation to imperialism), it arbitrarily and *inaccurately* relates this question *only* to industrial capital in the countries which annex other nations, and in an equally arbitrary and inaccurate manner brings out the annexation of agrarian regions.

Imperialism is a striving for annexations—this is what the *political* part of Kautsky's definition amounts to. It is correct, but very incomplete, for politically, imperialism is, in general, a striving towards violence and reaction. For the moment, however, we are interested in the *economic* aspect of the question, which Kautsky *himself* introduced into *his* definition. The inaccuracy of Kautsky's definition is strikingly obvious. The characteristic feature of imperialism is *not* industrial capital, *but* finance capital. It is not an accident that in France it was precisely the extraordinarily rapid development of *finance* capital, and the weakening of industrial capital, that, from 1880 onwards, gave rise to the extreme extension of annexationist (colonial) policy. The characteristic feature of imperialism is precisely that it strives to annex *not only* agricultural regions, but even highly industrialised regions (German appetite for Belgium; French appetite for Lorraine), because 1) the fact that the world is already divided up obliges those contemplating a *new* division to reach out for *any kind* of territory, and 2) because an essential feature of imperialism is the rivalry between a number of great powers in the striving for hegemony, i.e., for the conquest of territory, not so much directly for themselves as to weaken the adversary and undermine *his* hegemony. (Belgium is chiefly necessary to Germany as a base for operations against England; England needs Bagdad as a base for operations against Germany, etc.)

Kautsky refers especially—and repeatedly—to English writers who, he alleges, have given a purely political meaning to the word "imperialism" in the sense that Kautsky understands it. We take up the work by the Englishman Hobson, *Imperialism,* which appeared in 1902, and therein we read:

> The new imperialism differs from the older, first, in substituting for the ambition of a single growing empire the theory and the practice of competing empires, each motivated by similar lusts of political aggrandisement and commercial gain; secondly, in the dominance of financial or investing over mercantile interests.

We see, therefore, that Kautsky is absolutely wrong in referring to English writers generally (unless he meant the vulgar English imperialist writers, or the avowed apologists for imperialism). We see that Kautsky, wile claiming that he continues to defend Marxism, as a matter of fact takes a step backward compared with the *social-liberal* Hobson, who *more correctly* takes into account two "historically concrete" (Kautsky's definition is a mockery of historical concreteness) features of modern imperialism: 1) the competition between *several* imperialisms, and 2) the predominance of the financier over the merchant. If it were chiefly a question of the annexation of agrarian countries by industrial countries, the role of the merchant would be predominant.

Kautsky's definition is not only wrong and un-Marxian. It serves as a basis for a whole system of views which run counter to Marxian theory and Marxian practice all along the line. We shall refer to this again later. The argument about words which Kautsky raises as to whether the modern stage of capitalism should be called "imperialism" or "the stage of finance capital" is of no importance. Call it what you will, it matters little. The fact of the matter is that Kautsky detaches the politics of imperialism from its economics, speaks of annexations as being a policy "preferred" by finance capital, and opposes to it another bourgeois policy which, he alleges, is possible on this very basis of finance capital. According to his argument, monopolies in economics are compatible with non-monopolistic, non-violent, non-annexationist methods in politics. According to his argument, the territorial division of the world, which was completed precisely during the period of finance capital, and which constitutes the basis of the present peculiar forms of rivalry between the biggest capitalist states, is compatible with a non-imperialist policy. The result is a slurring-over and a blunting of the most profound contradictions of the latest stage of capitalism, instead of an exposure of their depth; the result is bourgeois reformism instead of Marxism.

Kautsky writes: "from the purely economic point of view it is not impossible that capitalism will yet go through a new phase: that of the extension of the policy of the cartels to foreign policy, the phase of ultra-imperialism," i.e., of a super-imperialism, a union of world imperialisms and not struggles among imperialisms: a phase of "the joint exploitation of the world by internationally combined finance capital."

We shall have to deal with this "theory of ultra-imperialism" later on in order to show in detail how definitely and utterly it departs from Marxism. In keeping with the plan of the present work, we shall examine the exact economic data on this question. Is "ultra-imperialism" possible "from the purely economic point of view" or is it ultra-nonsense?

If, by purely economic point of view a "pure" abstraction is meant, then all that can be said reduces itself to the following proposition: evolution is proceeding towards monopoly; therefore the trend is towards a single world monopoly, to a universal trust. This is indisputable, but it is also as completely meaningless as is the statement that "evolution is proceeding" towards the manufacture of foodstuffs in laboratories. In this sense the "theory" of ultra-imperialism is no less absurd than a "theory of ultra-agriculture" would be.

If, on the other hand, we are discussing the "purely economic" conditions of the epoch of finance capital as an historically concrete epoch which opened at the beginning of the twentieth century, then the best reply that one can make to the lifeless abstractions of "ultra-imperialism" . . . is to contrast them with the concrete economic realities of present-day world economy. Kautsky's utterly meaningless talk about ultra-imperialism encourages, among other things, that profoundly mistaken idea which only brings grist to the mill of the apologists of imperialism, *viz.*, that the rule of finance capital *lessens* the unevenness and contradictions inherent in world economy, where in reality it *increases* them.

R. Calwer, in his little book, *An Introduction to World Economics,* attempted to compile the main, purely economic, data required to understand in a concrete way the internal relations of world economy at the end of the nineteenth and beginning of the twentieth centuries. He divides the world into five "main economic areas," as follows: 1) Central Europe (the whole of Europe with the exception of Russia and Great Britain); 2) Great Britain; 3) Russia; 4) Eastern Asia; 5) America; he includes the colonies in the "areas" of the state to which they belong and "leaves out" a few countries not distributed according to areas, such as Persia, Afghanistan and Arabia in Asia; Morocco and Abyssinia in Africa, etc.

We notice three areas of highly developed capitalism with a high development of means of transport, of trade and of industry, the Central European, the British and the American areas. Among these are three states which dominate the world: Germany, Great Britain, the United States. Imperialist rivalry and the struggle between these countries have become very keen because Germany has only a restricted area and few colonies (the creation of "Central Europe" is still a matter for the future; it is being born in the midst of desperate struggles). For the moment the distinctive feature of Europe is political disintegration. In the British and American areas, on the other hand, political concentration is very highly developed, but there is a tremendous disparity between the immense colonies of the one and the insignificant colonies of the other. In the colonies, capitalism is only beginning to develop. The struggle for South America is becoming more and more acute.

There are two areas where capitalism is not strongly developed: Russia and Eastern Asia. In the former, the density of population is very low, in the latter it is very high; in the former political concentration is very high, in the latter it does not exist. The partition of China is only beginning, and the struggle between Japan, U.S.A., etc., in connection therewith is continually gaining in intensity.

Compare this reality, the vast diversity of economic and political conditions, the extreme disparity in the rate of development of the various countries, etc., and the violent struggles of the imperialist states, with Kautsky's silly little fable about "peaceful" ultra-imperialism. Is this not the reactionary attempt of a frightened philistine to hide from stern reality? Are not the international cartels which Kautsky imagines are the embryos of "ultra-imperialism" . . . an example of the division and the *redivision* of the world, the transition from peaceful division to non-peaceful division and *vice versa?* Is not American and other finance capital, which divided the whole world peacefully, with Germany's participation, for example, in the international rail syndicate, or in the international mercantile shipping trust, now engaged in *redividing* the world on the basis of a new relation of forces, which has been changed by methods *by no means* peaceful?

Finance capital and the trusts are increasing instead of diminishing the differences in the rate of development of the various parts of world economy. When the relation of forces is changed, how else, *under capitalism,* can the solution of contradictions be found, except by resorting to *violence?* Railway statistics provide remarkably exact data on the different rates of development of capitalism and finance capital in world economy. . . .

. . . . [T]he development of railways has been more rapid in the colonies and in the independent (and semi-dependent) states of Asia and America. Here, as we know, the finance capital of the four or five biggest capitalist states reigns undisputed. Two hundred thousand kilometers of new railways in the colonies and in the other countries of Asia and America represent more than 40,000,000,000 marks in capital, newly invested on particularly advantageous terms, with special guarantees of a good return and with profitable orders for steel works, etc., etc.

Capitalism is growing with the greatest rapidity in the colonies and in overseas countries. Among the latter, *new* imperialist powers are emerging (e.g., Japan). The struggle of world imperialism is becoming more acute. The tribute levied by finance capital on the most profitable colonial and overseas enterprises is increasing. In sharing out this "booty," an exceptionally large part goes to countries which, as far as the development of productive forces is concerned, do not always stand at the top of the list. . . .

. . . . [A]bout 80 per cent of the total existing railways are concentrated in the hands of the five Great Powers. But the concentration of the *ownership* of these railways, of finance capital is much greater still: French and English millionaires, for example, own an enormous amount of stocks and bonds in American, Russian and other railways.

Thanks to her colonies, Great Britain has increased the length of "her" railways by 100,000 kilometers, four times as much as Germany. And yet, it is well known that the development of productive forces in Germany, and especially the development of the coal and iron industries, has been much more rapid during this period than in England—not to mention France and Russia. In 1892, Germany produced 4,900,000 tons of pig iron and Great Britain produced 6,800,000 tons; in 1912, Germany produced 17,600,000 tons and Great Britain 9,000,000 tons. Germany, therefore, had an overwhelming superiority over England in this respect. We ask, is there *under capitalism* any means of removing the disparity between the development of productive forces and the accumulation of capital on the one side, and the division of colonies and "spheres of influence" for finance capital on the other side— other than by resorting to war?

4

Revisions of Liberalism: Interdependence Theory

Modern liberalism has both affirmed and denied the theoretical project of international political economy. Neoclassical analysis essentially ignores the problem of how economics and politics interact, focusing exclusively on "pure economics." On the other hand, John Maynard Keynes, with his ideas of macroeconomic aggregates and countercyclical state intervention, lays the analytic groundwork for a revised liberal IPE theory. Although Keynes himself did not fully develop the political side of the argument, his ideas have been incorporated and extended by interdependence theorists.

Keynesianism resonates in the work of Richard Cooper, one of the first to popularize the notion of interdependence. Cooper suggests that advances in transportation, communications, technology transfer, and the like have fundamentally altered the conditions of international economics. Most importantly, differences in comparative costs appear to be diminishing. This implies that the gains from trade seen by Smith and Ricardo may not be as obvious as they once were. Furthermore, global economic integration may thwart equilibrating world market forces: "in normal periods prospective imbalances in international payments . . . are likely to be more frequent and of larger amplitude than they have been in the past." In a manner similar to Keynes's discussion of the difficulties of achieving domestic equilibrium, Cooper argues that new policies are needed to address the unprecedented conditions of international interdependence.

Although diminishing comparative cost differentials in an interdependent world may tempt national leaders to return to mercantilist practice, Cooper vigorously opposes such a course. Invoking Keynes, he argues that economic nationalism invites policy competition that is doomed to fail: "like members of a crowd rising to their tip-toes to see a parade better but in the end merely standing uncomfortably on their tip-toes." National economic independence is further constrained by contemporary international conventions, such as the General Agreement on Tariffs and Trade, that limit policy instruments (e.g., quantitative trade restrictions) available to state leaders. Cooper concludes that international policy coordination is virtually the only means to achieve national economic goals in an interdependent world.

Cooper thus defines the challenges of interdependence and prescribes cooperation as the best response. He, again like Keynes, does not consider how political conditions may promote or undermine international cooperation. Presumably, state leaders will realize the irrationalities of policy competition and state interests will come to be defined in terms of cooperation. Robert Keohane and Joseph Nye go a step further and analyze how international politics is transformed by interdependence.

Keohane and Nye are engaged in a conversation with international relations realists. Out of their critique of realism, they construct an ideal typical vision of IPE, complex interdependence, that falls within the purview of the liberal tradition. Realism, they suggest, is based on two key assumptions: the preeminence of states as world actors and the centrality of military force as international power. Keohane and Nye argue that global economic interdependence has cast doubt upon these assumptions. Transnational corporations and organizations born of economic integration now vie with states for global influence. Military power is essentially useless in a wide variety of significant issue-areas; firepower has little relevance in international monetary negotiations among regional allies. Moreover, the advent of nuclear weapons has dramatically raised the cost of employing military force, reducing its utility. Traditional realism, Keohane and Nye contend, does not take account of these changing world conditions.

In situations of interdependence, new power relationships are established, based upon mutual asymmetrical dependencies. Keohane and Nye hold that the politics of interdependence are best understood in terms of sensitivity and vulnerability. Sensitivity applies when costs imposed from without can be minimized by a change in existing policy. If costs cannot be significantly reduced even with a change in policy, then the situation is one of vulnerability. The costliness of alternative courses of action will vary from issue area to issue area; a state may be vulnerable to sudden changes in world oil prices, but be only sensitive to fluctuations in the supply of certain strategic minerals. Moreover, within a given issue area, one actor may be sensitive to changing conditions while others are vulnerable. In political terms, vulnerabilities are most important, as they may provide sources of power that realism fails to elucidate.

The politics of interdependence is also influenced by international regimes. Keohane and Nye define regimes as "networks of rules, norms, and procedures that regularize behavior" in a given issue-area. Regimes are affected by the distribution of power among states, but regimes, in turn, may critically influence the bargaining process among states. The point here is that realist analysis of systemic balances of power is insufficient to comprehend the modern world. Moreover, state interests may be constrained by regimes. In some instances, regimes are the "rules of the game" that limit even the most powerful actors. International regimes are therefore important loci of the politics of interdependence. Keohane and Nye are modest in generalizing about regime creation and maintenance, pointing out that further research is necessary to explain such processes. They offer a prolegomenon to interdependence theory that inspired many students of IPE in the 1970s and 1980s.

Keohane and Nye agree with Cooper that interdependence has transformed international relations. They also agree that interdependence offers new opportunities for cooperation. States exposed to interdependence vulnerabilities may seek compromises on many issues. Keohane and Nye, however, argue against an overly optimistic attitude toward global harmony, which they label the "modernist" position. Although they are somewhat more circumspect toward potential cooperation, they are, in the final analysis, consistent with general liberal orientations. Their interpretation of interdependence assumes a world in which politics, while not wholly transformed, is fundamentally shaped by economic forces.

National Economic Policy
in an Interdependent World Economy

RICHARD COOPER

During the past decade [1960s] there has been a strong trend toward economic interdependence among the industrial countries. This growing interdependence makes the successful pursuit of national economic objectives much more difficult. Broadly speaking, increasing interdependence complicates the successful pursuit of national economic objectives in three ways. First, it increases the number and magnitude of the distrubances to which each country's balance of payments is subjected, and this in turn diverts policy attention and instruments of policy to the restoration of external balance. Second, it slows down the process by which national authorities, each acting on its own, are able to reach domestic objectives. Third, the response to greater integration can involve the community of nations in counteracting motions which leave all countries worse off than they need be. These difficulties are in turn complicated by the fact that the objective of greater economic integration involves international agreements which reduce the number of policy instruments available to national authorities for pursuit of their economic objectives. This chapter touches on all of these facets of higher economic interdependence among industrial nations, both as fact and as objective, but its principal focus is on the third complication—the process of mutually damaging competition among national policies.

. . . . [B]oth institutional and economic changes have increased economic interdependence among the industrial countries since the late 1940s. Import quotas in industrial countries have been virtually abolished on trade in manu-

Reprinted from *The Economics of Interdependence: Economic Policy in the Atlantic Community* by Richard Cooper, "National Economic Policy in an Interdependent World Economy," by permission of McGraw-Hill Publishing Company.

facture products; tariffs have been reduced; and transportation costs have fallen relative to the value of goods. At the same time, the accumulation of capital and the spread of technology have made national economies more similar in their basic characteristics of production; comparative cost differences have apparently narrowed, suggesting that imports can be replaced by domestic production with less loss in national income than heretofore. Whether a country imports a particular good or exports it thus becomes less dependent on the basic characteristics of the economy, more dependent on historical development and on relatively accidental and transitory features of recent investment decisions at home and abroad. An invention in one country may lead to production there for export, but the new product will relatively quickly be produced abroad—or supplanted by a still newer product—and possibly even exported to the original innovating country.

Enlargement of the decision-making domain of the world's great producing firms results in the rapid movement of capital and technical knowledge across national frontiers, thereby contributing to the narrowing of comparative cost differences; but their activity will also quicken the speed with which trade adjusts to new sales opportunities because they have direct knowledge of foreign markets and access to distribution channels.

Monetary disturbances, too, are likely to be much more quickly translated into changes in the volume of exports and imports than they were formerly. Under fixed exchange rates, greater than average monetary inflation in one country will invite a more rapid deterioration in the balance on goods and services than was true in the past.

Finally, as financial markets become more closely integrated, relatively small differences in yields on securities will induce large flows of funds between countries. Banks will increasingly number "foreign" firms among their prime customers; the advantages of inexpensive credit to firms in countries with ample savings and well-functioning financial markets, such as the United States, will be shared increasingly with firms elsewhere.

All these changes in the characteristics of the international economy during the past decade—and it should be emphasized again that economic integration is still far from complete—are crucial to the functioning of the international payments system and the autonomy which it permits in the formation of national economic policy. These changes mean that in normal periods prospective imbalances in international payments—imbalances that would arise if countries did not respond to reduce them or did not adjust policy measures to forestall them—are likely to be more frequent and of larger amplitude than they have been in the past. "Disturbances" arising from new innovations, from generous wage settlements leading to price increases, and from excess or deficient domestic demand will affect the balance of payments more perceptibly. Whether or not imbalances also last longer depends upon the relationship among the "disturbances"; if they are well distributed among countries and tend equally toward deficit or surplus, the duration of imbalances may well be less than in the past; otherwise it may be longer.

These changes suggest that prospective balance-of-payments difficulties are likely to be more common, and that they will worsen as the structural changes continue in their recent trend. By the same token, however, correction of imbalances in international payments should be easier in the future. Trade flows will respond more sharply to small "disturbances;" but the flows should also respond more quickly to policy measures designed to influence them. If a small relative increase in the price level will lead it into greater balance-of-payments difficulties than before, a relatively small decrease should undo the difficulties. Similarly, international capital flows will respond more readily to small differences in national credit conditions; but small differences in national credit conditions directed to correcting the imbalance can induce equilibrating flows of capital. Thus if the national authorities can recognize disturbances early, are willing to use some of the tools at their disposal for correcting imbalances in international payments, and can act reasonably quickly in doing so, then the increased sensitivity of payments to various disturbances need cause no undue difficulty, provided that policy instruments are properly chosen and adequately coordinated among countries.

Interdependence before 1914

There is a natural inclination to compare the international economy of today, especially in the claim that it is becoming more integrated, with the international economy before 1914, when, it is often said, the world economy was highly integrated. In the four decades before World War I, most of the major countries were customarily on the gold standard (implying fixed exchange rates), capital was free to move into or out of most countries, trade was impeded only by comparatively moderate tariffs, and quotas were generally absent. Even labor was generally free to migrate from country to country without visas, security checks, and immigration quotas.

In one important sense, however, the comparison is not at all apt. Today national governments are much more ambitious about the objectives of national economic policy than they were in the nineteenth century. . . . Governments have taken on the responsibility for assuring high levels of employment and, increasingly, a rapid rate of growth; and they attempt actively to influence the allocation of resources and the distribution of income to a much greater degree. These new tasks place greater burdens on the available instruments of policy. Before 1914, by contrast, preoccupation with "defending the currency" was dominant, and the (admittedly more limited) policy instruments at hand were more willingly devoted to that end. Thus, the intrusions of international economic integration on national economic policy was more readily accepted because national economic policy was far less ambitious in its aims.

In addition to this important difference, economic relations among industrial countries are probably potentially much closer today than they were even before 1914, despite the characteristics of the pre-1914 world noted above. True, British and French capital moved overseas readily and British investors built railroads around the world. The proportion of Britain's annual savings which went abroad was, in fact, staggering by modern standards. Nonetheless, communications were far less perfect than they were today and foreign investors ran greater commercial risks arising from imperfect knowledge (except in the case of colonial bonds that in effect had the sponsorship of the home government).

Despite the freedom of capital to move, it did not in fact move in sufficient volume even to erase differences in short-term interest rates. Over the period 1876–1914, short-term interest rates in New York averaged more than one percentage point higher than corresponding rates in London, and there was only a weak correspondence in movement between short-term rates in the two financial centers. Short-term interest rates in London and Paris were much closer together and the correspondence in their movement was higher but still far from perfect. Long-term interest rates showed similar divergence in their levels and movement. Response to new investment opportunities, when it came at all, was often slow.

While tariffs were generally low, barriers to trade in the form of transportation costs were very substantial, although they declined sharply after the introduction of the ocean steamship. Large differences in comparative costs meant trade was socially very profitable, but the composition and level of trade was correspondingly less sensitive to small changes in costs, prices, and quality. Finally, business organizations, far from being international, became truly national corporations in the United States only with the approach of World War I, and the process was even slower in many European countries.

Thus, the integration of the pre-1914 world economy was something of an illusion. While the pre-1914 world was integrated in the sense that government-imposed barriers to the movement of goods, capital and people were minimal, those imposed by nature were much greater and economic integration was not high in the sense used here: quick responsiveness to differential earning opportunities resulting in a sharp reduction in differences in factor rewards.

Countries today are gradually entering a new environment, not merely returning to a condition that once existed. They confront new problems arising from the combination of more ambitious national and international economic objectives and a potentially higher degree of economic interdependence than has ever existed before. . . . It is now necessary to specify more precisely how conflicts may arise and to indicate some of the ways in which governments have responded to those conflicts.

Economic Objectives and Policy Instruments

A well known proposition in the theory of economic policy requires that the number of policy instruments be at least as great as the number of objectives (target variables) if all objectives are to be achieved. If the number of instruments is fewer than the number of targets, it will not be possible to reach all of the targets; in that case at least some targets must be given up, and the authorities must choose among them.

A simple example can illustrate the need to have at least as many instruments as targets. Suppose the government of an isolated country has two economic objectives: it would like to assure full employment of its labor forces at all times, and it would like its national product to grow at a specified rate each year. It can vary the over-all size of the budget deficit or surplus (fiscal policy) to assure full employment. But full employment of resources can be met with a variety of combinations of investment, consumption, and government expenditure. Without some other instrument, the desired growth rate cannot be assured. If, however, investment is stimulated by a low rate of interest and higher investment leads to more growth, then monetary policy and fiscal policy together can be manipulated to achieve the two objectives. The higher the growth rate desired, the lower should be the rate of interest. Fiscal policy can then be adjusted to assure full employment. This very simple model apparently influenced thinking in the early years of the Kennedy administration.

Viewing economic policy as a problem in specifying targets and finding sufficient instruments to reach them helps to illuminate many policy problems confronting national authorities . . . The objective of greater economic integration has led many officials to reject both flexible exchange rates and frequent variations in fixed exchange rates as instruments for maintaining balance-of-payments equilibrium. A number of other instruments of policy have been ruled out by international agreement on the same grounds, or on the grounds that their use was likely to lead to retaliation and counter-retaliation that would leave countries worse off than they were at the outset. Most types of export subsidy, tariff discrimination among countries, increases in tariffs, and discriminatory exchange regulations fall into this category. A number of provisions of the GATT are devoted to these exclusions and prohibitions; with specified exceptions, such as the formation of customs unions or free trade areas, trade discrimination is proscribed, as are many types of export subsidies and discrimination in domestic taxation between home and foreign goods. The IMF Articles of Agreement make similar prohibitions with respect to currency arrangements. . . . The extensive use of these measures in the past, especially in the 1930s, led to widespread retaliation and mutual recriminations, and they acquired a bad name among outward looking officials. But the price of international rules of good behavior as set forth in the GATT and the IMF Articles has been a reduction in the range of instrument available to national policy-makers.

Some policy instruments may be used, as a practical matter, only within a limited range. In the United States, changes in the discount rate of the Federal Reserve System and (since 1962) deliberate deficits or surpluses in the government budget are both regarded as legitimate tools of economic policy; but in normal times the public is not likely to countenance a discount rate of 20 per cent or a budget deficit of $50 billion. These exceed the range of acceptability; policy instruments have "boundary conditions." In the abnormal situations when such limits become operative, an instrument is withdrawn from use.

Sometimes these limits are not fully known until they are tested; then we discover that we have more targets (or fewer instruments) than were previously apparent.

It goes without saying that to be attainable, economic objectives must be consistent. If they are not consistent, no number of policy instruments will suffice to reach the objectives. One illustration in the forefront of discussion in most industrial countries involves the relationship between employment and price stability. Given the institution of private collective bargaining, is the target of "full employment" (4 per cent unemployment in the United States, under 2 per cent in the United Kingdom, according to the standards and definitions accepted by each) consistent with "price stability," defined, say, as stability in the consumer price index? Many economists would find a conflict.

This kind of inconsistency can perhaps be overcome by developing new policy instruments. Another kind of inconsistency, especially important to national economies linked through international trade and capital movements, cannot be eliminated through the development of new instruments. Examples are objectives regarding the balance of payments and the trade balance. Since one country's trade surplus is another country's trade deficit, it is impossible for all countries to succeed in running trade surpluses. The same is true for balance of payments, taking into account capital movements. If there are n countries, only $n-1$ of them can succeed in reaching their independent balance-of-payments targets; at least one must accept defeat or else fail to target values for its trade position and its balance-of-payments position, thereby acting as an international residual. It has been suggested that the United States played this role until the late 1950s, by taking a relatively passive position toward its payments position after the termination of Marshall Plan aid.

The requirement of consistency is not merely theoretical. In 1962, for instance, all of the major industrial countries wanted simultaneously to improve their payments positions on current account. While mutual success was logically possible in this case, it did imply a correspondingly sharp deterioration in the current account position of the less developed countries taken together, which in turn would require ample financing from the industrial countries in the form of grants or loans. No such increase in capital outflows was targeted. Thus, national targets were inconsistent.

The Speed of Adjustment

In summary, successful economic policy requires an adequate number of policy instruments for the number of economic objectives, and it requires that these objectives be consistent with one another. If either of these conditions fails, policy-makers are bound to be frustrated in their efforts. Before turning to how these frustrations become manifest, however, one further point should be made: growing interdependence can slow down greatly the process by which independently acting national authorities reach their economic objectives, even when all the targets are consistent and there are sufficient policy instruments at hand to reach them. Thus, in practice, nations may find themselves farther from their objectives than would be true with less independence.

High interdependence slows the speed of adjustment to disturbances if national policy-makers do not take the interdependence into account. This is because the economic authorities in one country may be working at cross purposes with those in another. An investment boom in one country may raise interest rates both at home and, by attracting internationally mobile funds, in neighboring countries. The first country may temporarily welcome the high interest rates to help curb the boom and may also tighten fiscal policy to keep inflationary pressures in check. But other countries may fear that higher interest rates will deter investment at home and take steps to lower interest rates. Unless this monetary relaxation is taken into account in framing fiscal policy in the first country, its authorities will find that fiscal policy has not been sufficiently contractionary. But more contractionary fiscal policy will tend to hold interest rates up, so that the monetary authorities in the neighboring countries will find they have only been partially successful in lowering their rates. Even if in the end the whole process settles to a point where the various national authorities are satisfied, it will have taken longer than if there had been close coordination between the authorities in the several countries involved. The greater the interaction between the countries, the longer convergence will take if countries act solely on their own.

If policy decisions are truly decentralized among nations, in the sense that the authorities in each nation pursue only their own objectives with their own instruments without taking into account the interactions with other countries, then the more interdependent the international economy is, the less successful countries are likely to be in reaching and maintaining their economic objectives. This is due to the greater impact of domestic measures on foreign economies, calling forth correspondingly greater offsetting responses which in turn affect the first country. Under these circumstances, countries must either reconcile themselves to prolonged delays in reaching their objectives or they must coordinate their policies more closely with those of other nations.

International Competition in Economic Policy

In an interdependent economy, governments do not have full control over the instrument variables needed to influence the trade balance or the balance of payments. Each government can affect the domestic interest rate in an attempt to influence international capital movements or can set tariffs on imports and subsidies on exports to influence the trade balance. But success in influencing capital movements or trade flows depends on what other countries are doing. It is interest rate *differentials,* not the absolute level of interest rates, which induce the movement of capital. And it is domestic tariffs *less* foreign subsidies which influence the level of imports. There are many instruments of economic policy for which relative differences affect international transactions, but the absolute value may continue to exert a strong influence on purely domestic decisions. This is true, for example, not only of short- and long-term interest rates, but also of liberal tax benefits to investment, generous depreciation allowances, lax regulation of corporate activities and a host of other measures designed to influence corporate location. It is also true of foreign trade: generous credit arrangements or credit-risk guarantees for exports may encourage total exports without improving the trade balance if other countries are pursuing similar measures.

This feature of policy instruments—that the absolute level of the instrument may have important effects domestically, but that only the level relative to that in other countries influence the balance of trade or payments—raises the question: where do the values of these instruments finally settle? International capital movements between two otherwise isolated countries will presumably be roughly the same whether interest rates are at 7 per cent in one and 5 per cent in the other or at 4 per cent in the first country and 2 per cent in the second. In each case, the differential is two percentage points. But what determines whether "community" interest rates settle at the higher level or the lower one? The effects on other objectives may be very different. Economic growth will be inhibited more in the first case than in the second.

The values that policy instruments take on in the community of nations, and the process by which those values are reached, are of strong interest to the individual nations. They may not have sufficient domestic flexibility to offset the damaging effects of policy instruments that are forced to an inappropriate level by international competition among governments. As a result, greater international integration can force choices among national objectives, all of which might otherwise be attainable.

There are situations in which most or even all members of the international community will find themselves worse off. The competitive devaluations and tariff wars of the interwar period offer the most striking example; many of the proscriptions in the GATT and the IMF Articles of Agreement are designed to avoid a repetition of those events.

But competition among policies was not thereby banished on all fronts. For example, interest rates shot upward in 1965 and 1966 to levels one to two percentage points higher than those which had prevailed in most countries in 1964. Some of the increases were designed to curb domestic demand; others were defensive, to limit capital outflow. Even after domestic economies had cooled down, it took a dramatic meeting of finance ministers at Checquers, England, in early 1967, to reverse the process.

Four other types of policy instruments having these characteristics have been used in the effort to strengthen the balance of payments of various countries: restrictions on government procurement, government-sponsored export promotion, tax incentives to domestic investment, and changes in domestic tax structure. The United States, faced with large payments deficits during the early 1960s, made or considered moves in all of these areas. In each case there was ample precedent abroad for doing so.

Government purchases for government use are specifically excluded from coverage by the GATT rules governing international trade. The result is that a conspicuously small proportion of government purchases, by any government, is from foreign suppliers who compete with domestic producers. In the United States the "Buy American" provision, which after 1954 officially gave preferential treatment of 6 to 12 per cent (in addition to tariffs) to domestic over foreign competitors, has existed since the 1930s. But in 1962, a number of government agencies, including most importantly the Department of Defense, raised the preference accorded to domestic supplies as high as 50 per cent. Foreign aid expenditures by the American government are even more restricted. Starting with development loans in 1959, such expenditures were tied increasingly to purchases in the United States, until only a limited class of expenditures was not so tied, regardless of the price advantages offered by foreign suppliers.

The government procurement practices of other countries are more difficult to document, since most governments do not require open bidding on government purchases with well-publicized preferences to domestic producers, such as are found in the "Buy American" provisions. Many countries follow the practice of tying foreign assistance, either by law or by skillful selection of projects and recipient countries, to purchases from the donor country. This is as true for those donors with fully employed economies as for those with excess capacity and unemployment, even though tying is far less effective in the former case, and merely stimulates additional imports; and it is as true for donor countries in balance-of-payments surplus as for those in deficit. . . .

Many of these practices, of course, arise not only from balance-of-payments considerations but also from protectionist sentiment. Domestic producers apply strong political pressures on their governments to buy at home, especially when the goods are to be "given away." But weakness in the balance of payments often strengthens their arguments and increases public acceptibility of such restrictive measures.

Government activities are not solely restrictive of trade. On the contrary, a second range of practices involves all kinds of schemes, except direct subsidies proscribed by GATT, to promote exports of goods and services. Governments sponsor trade fairs, product exhibitions, and other advertisements for the products of their exporters; they insure commercial and so-called noncommercial risks involved in exporting; and they often help to finance exports directly. No major industrial trading nation can be found without a government or government-sponsored agency for insuring and/or extending credit for exports. Some countries, such as France and Italy, give especially favorable treatment to export paper in their banking systems or at their central banks. Export credit is often exempt from general credit limitations to restrict domestic demand. All of these measures really subsidize exports, although it is often impossible to identify the amount of the subsidy in any particular sale.

Subsidies to domestic investment is another area in which governments have moved to improve their international payments positions. Investment subsidies for manufacturing and agriculture improve the competitiveness of a country's products in world markets. Some countries give direct fiscal incentives to new investment in plant and equipment, such as the investment tax credit of 7 per cent adopted by the United States in 1962 and the 25 per cent investment grants in the United Kingdom. . . .

Under a regime of fixed exchange rates, government subsidy for domestic investment is similar to a devaluation of the currency in that it improves the cost competitiveness both of the country's export products and of its products which compete with imports.

Subsidies to investment are obviously motivated by considerations extending well beyond the balance of payments; economic growth has become a target of economic policy in its own right, partly for political and strategic reasons (arising in part from the "economic race" with the Soviet bloc), partly because rising standards of living are universally desired. But balance-of-payments considerations do play an important role in the decision to inaugurate investment incentives. Britain for years has emphasized the need to enlarge and improve its capital stock to compete more effectively in world markets. . . .

Changes in the structure of domestic taxation, and in particular the "mix" between direct and indirect taxes, constitutes another area in which governments have moved, or have been tempted to move, to improve their national trade positions. GATT rules prohibiting export subsidies have been interpreted to preclude remission of direct taxes on exports but to permit remission of indirect taxes. Thus taxes on corporate profits arising from export cannot be rebated, but manufacturers' excise taxes or turnover taxes can be. Similarly, countries are permitted to levy indirect taxes, but not direct taxes, on imports. Because of this asymmetry in border tax adjustment, it is possible under fixed exchange rates for a country to stimulate exports and to impede imports by shifting its tax structure from direct taxes to indirect taxes, provided that direct taxes affect prices.

All of these policy measures have a common characteristic. Taken by one country alone, each represents a concealed devaluation of the currency, at least with respect to a selected class of transactions. But like devaluation, these measures are effective only if other countries do not respond in kind. To each country, tying foreign aid and giving preference to domestic producers in government procurement may appear to offer a means to improve the balance of payments; indeed, in the short run it may do so. But if all countries follow the same practices,the benefit to each is much reduced and some countries will have their payments positions worsened as a result. In the meantime, the total real value of foreign aid has been reduced by reliance on high cost suppliers, and inefficient production has been fostered.

The same thing is true of the other measures discussed. General adoption of export promotion schemes and government-sponsored tourist publicity will surely have a much greater effect on the total level of world exports and tourism than on the payments position of any one country, since the measures will largely cancel one another and leave only residual effects on the balance of payments. Similarly, if all countries adopt special tax incentives for domestic investment, the net improvement in competitiveness—which depends as much on incentives abroad as on those at home—will be haphazard and unpredictable. The principal effect may well be not on any one country's balance-of-payments position but on the total investment and the rate of growth in the world economy at large—so long as these effects are not nullified by a competitive rise in long-term interest rates! Finally, an effort to raise exports and impede imports through changes in domestic tax structure may have little over-all effect on foreign trade and leave countries with tax structures which many would prefer not to have.

At any point in time, there are often cogent and persuasive arguments for introducing one or more of these measures to improve the balance of payments. If other countries did not respond in kind, the desired improvement would be forthcoming. But if other countries act likewise, the measures largely cancel out. Not only is the purpose of the move nullified, but all countries may find themselves worse off in terms of their other objectives. As a rule, individual countries cannot act unilaterally without inviting reaction. If they are successful, they are quickly emulated by their neighbors, so that the initial gains are transitory at best. Countries often must act in self-defense, in response to the behavior of their trading partners. This is particularly so when measures to reduce one country's deficit do not reduce the surpluses of the surplus countries but increase the deficit of another deficit country or move countries in balance into deficit. These third countries then feel compelled to respond defensively and their actions in turn increase the deficit of the initial deficit country. Moreover, many of the measures thus taken are difficult to reverse; countries do not readily contract export credit programs or lengthen the period of depreciation allowable for tax purposes.

In Summary

Contemporary competition among policies is not obvious, as it was in the round of tariff increases in the late 1920s and the competitive depreciations of the early 1930s. But more subtle and sophisticated methods can substitute, albeit imperfectly, for currency depreciation. Taken in sequence by different countries, these measures produce a kind of ratchet effect. We then have a series of competitive depreciations in disguise.

This chapter has focused on how balance-of-payments difficulties, actual or feared, can give rise to undesirable competition in policies. . . .

These developments are understandable, and can be expected to become more common. In a highly integrated economic area which surpasses in size the jurisdiction of governments, each group of policy-makers is subject to such strong interactions with the surrounding area that the constraints on its actions become very severe. Indeed, in the hypothetically limiting case, these constraints determine entirely the course of action each jurisdiction must take. The region, or the nation, in a highly integrated economy becomes analogous to the perfect competitor—or at best the oligopolist—in market economy. The range of choice it has, consistent with economic survival, is very small; for the most part it simply adapts its behavior to stimuli from outside. Awareness of the high interactions will eventually inhibit action.

A. C. Pigou and John Maynard Keynes pointed out long ago that the sum of individual decisions by consumers and producers may not always be optimal for society as a whole (and hence for its members), even though its members may be acting individually on entirely rational grounds. Some kind of collective action is therefore required to produce an optimal outcome.

The same can be true among nations or among regions within a nation, if the interactions among their decisions are sufficiently strong. One jurisdiction gropes around for new instruments in an attempt to improve its position. If it succeeds, others follow and there is a competition in policies which defeats everyone's objectives and in fact can even lead all participants *away* from their national or local objectives, like the members of a crowd rising to their tip-toes to see a parade better but in the end merely standing uncomfortably on their tip-toes.

An invisible hand seems to be working in the area of economic policy as well as in the market place. Competition in the market place is alleged to lead to the most efficient allocation of resources. Whatever the merits of this claim, we can be much less confident that competition among policies will be optimal. Governments seek many ends, not the efficient allocation of resources alone; and the process of policy competition can certainly thwart some of those objectives.

Existing rules of international behavior as set forth in GATT and in the Bretton Woods Agreement do limit the use of direct and straightforward means of policy competition such as open export subsidies and multiple exchange rates, and they therefore slow the process of policy competition since the more subtle and sophisticated methods—loopholes in GATT and the Bretton Woods Agreement—usually involve strong domestic considerations which delay their implementation. But existing rules do not fully accomplish the aim of preventing self-defeating policy competition and of freeing domestic policy measures to purse largely domestic objectives. Moreover, the pressures on domestic policy are likely to become greater as the world economy becomes more interdependent. Freedom of action in economic policy formation can be lost through the need for each country to compete in policies with its competitors in commerce.

. . . . Countries can coordinate closely their national economic policies, attempting to define and reach an optimum combination of policies for the community as a whole. This route involves extensive "internationalization" of the process of economic policy-making, transferring this government function to the larger integrated area.

Alternatively, countries can attempt to remove the major source of pressure on their actions—their unfavorable international payments positions—by providing each country with ample liquidity to finance any deficit and allowing it to go its own way. Or this goal can be achieved by reversing the growth in interdependence, by artificially breaking down or reducing the numerous economic links between countries. . . .

Interdependence in World Politics

ROBERT O. KEOHANE AND JOSEPH S. NYE

We live in an era of interdependence. This vague phrase expresses a poorly understood but widespread feeling that the very nature of world politics is changing. The power of nations—that age-old touchstone of analysts and statesmen—has become more elusive. . . .

How profound are the changes? A modernist school sees telecommunications and jet travel as creating a "global village" and believes that burgeoning social and economic transactions are creating a "world without borders." To greater or lesser extent, a number of scholars see our era as one in which the territorial state, which has been dominant in world politics for the four centuries since feudal times ended, is being eclipsed by nonterritorial actors such as multinational corporations, transnational social movements, and international organizations. As one economist put it, "the state is about through as an economic unit."

Traditionalists call these assertions unfounded "globaloney." They point to the continuity in world politics. Military interdependence has always existed, and military power is still important in world politics—witness nuclear deterrence; the Vietnam, Middle East, and India-Pakistan wars; and Soviet influence in Eastern Europe or American influence in the Caribbean. Moreover, as the Soviet Union has shown, authoritarian states can, to a considerable extent, control telecommunications and social transactions that they consider disruptive. Even poor and weak countries have been able to nationalize multinational corporations, and the prevalence of nationalism casts doubt on the proposition that the nation-state is fading away.

Neither the modernists nor the traditionalists have an adequate framework for understanding the politics of global interdependence. Modernists point correctly to the fundamental changes now taking place, but they often assume without sufficient analysis that advances in technology and increases in social and economic transactions will lead to a new world in which states, and their control of force, will no longer be important. Traditionalists are adept at showing flaws in the modernist vision by pointing out how military interdependence continues, but find it very difficult accurately to interpret today's multidimensional economic, social, and ecological interdependence.

Our task . . . is not to argue either the modernist or traditionalist position. Because our era is marked by both continuity and change, this would be fruitless. Rather, our task is to provide a means of distilling and blending the wisdom in both positions by developing a coherent theoretical framework for the political analysis of interdependence. We shall develop several different but potentially complementary models, or intellectual tools, for grasping the reality of interdependence in contemporary world politics. Equally important, we shall attempt to explore the *conditions* under which each model will be most likely to produce accurate predictions and satisfactory explanations. . . .

Interdependence As An Analytic Concept

In common parlance, *dependence* means a state of being determined or significantly affected by external forces. *Interdependence,* most simply defined, means *mutual* dependence. Interdependence in world politics refers to situations characterized by reciprocal effects among countries or among actors in different countries.

These effects often result from international transactions—flows of money, goods, people, and messages across international boundaries. Such transactions have increased dramatically since World War II: Yet this interconnectedness is not the same as interdependence. The effects of transactions on interdependence will depend on the constraints, or costs, associated with them. A country that imports all of its oil is likely to be more dependent on a continual flow of petroleum than a country importing furs, jewelry, and perfume (even of equivalent monetary value) will be on uninterrupted access to these luxury goods. Where there are reciprocal (although not necessarily symmetrical) costly effects of transactions, there is interdependence. Where interactions do not have significant costly effects, there is simply interconnectedness. The distinction is vital if we are to understand the *politics* of interdependence.

Costly effects may be imposed directly and intentionally by another actor—as in Soviet–American strategic interdependence, which derives from the mutual threat of nuclear destruction. But some costly effects do not come directly or intentionally from other actors. For example, collective action may be necessary to prevent disaster for an alliance (the members of which are interdependent), for an international economic system (which may face chaos because of the absence of coordination, rather than through the malevolence of any actor), or for an ecological system threatened by a gradual increase of industrial effluent.

We do not limit the term *interdependence* to situations of mutual benefit. Such a definition would assume that the concept is only useful analytically where the modernist view of the world prevails: where threats of military force are few and levels of conflict are low. It would exclude from interdependence cases of mutual dependence, such as the strategic interdependence between the United States and the Soviet Union. Furthermore, it would make it very ambiguous whether relations between industrialized countries and less developed countries should be considered interdependent or not. Their inclusion would depend on an inherently subjective judgment about whether the relationships were "mutually beneficial."

Because we wish to avoid sterile arguments about whether a given set of relationships is characterized by interdependence or not, and because we seek to use the concept of interdependence to integrate rather than further to divide modernist and traditional approaches, we choose a broader definition. Our perspective implies that interdependent relationships will always involve costs, since interdependence restricts autonomy; but it is impossible to specify *a priori* whether the benefits of a relationship will exceed the costs. This will depend on the values of the actors as well as on the nature of the relationship. Nothing guarantees that relationships that we designate as "interdependent" will be characterized by mutual benefits.

Two different perspectives can be adopted for analyzing the costs and benefits of an interdependent relationship. The first focuses on the joint gains or joint losses to the parties involved. The other stresses *relative* gains and distributional issues. Classical economists adopted the first approach in formulating their powerful insight about comparative advantage: that undistorted international trade will provide overall net benefits. Unfortunately, an exclusive focus on joint gain may obscure the second key issue: how those gains are divided. Many of the crucial political issues of interdependence revolve around the old question of politics, "who gets what?"

It is important to guard against the assumption that measures that increase joint gain from a relationship will somehow be free of distributional conflict. Governments and nongovernmental organizations will strive to increase their shares of gains from transactions, even when they both profit enormously from the relationship. Oil-exporting governments and multinational oil companies, for instance, share an interest in high prices for petroleum; but they have also been in conflict over shares of the profits involved.

We must therefore be cautious about the prospect that rising interdependence is creating a brave new world of cooperation to replace the bad old world of international conflict. As every parent of small children knows, baking a larger pie does not stop disputes over the size of the slices. An optimistic approach would overlook the uses of economic and even ecological interdependence in competitive international politics.

The difference between traditional international politics and the politics of economic and ecological interdependence is *not* the difference between a world of "zero–sum" (where one side's gain is the other side's loss) and "nonzero–sum" games. Military interdependence need not be zero–sum. Indeed, military allies actively seek interdependence to provide enhanced security for all. Even balance of power situations need not be zero–sum. If one side seeks to upset the status quo, then its gain is at the expense of the other. But if most or all participants want a stable status quo, they can jointly gain by preserving the balance of power among them. Conversely, the politics of economic and ecological interdependence involve competition even when large net benefits can be expected from cooperation. There are important continuities, as well as marked differences, between the traditional politics of military security and the politics of economic and ecological interdependence.

We must also be careful not to define interdependence entirely in terms of situations of *evenly balanced* mutual dependence. It is *asymmetries* in dependence that are most likely to provide sources of influence for actors in their dealings with one another. Less dependent actors can often use the interdependent relationship as a source of power in bargaining over an issue and perhaps to affect other issues. At the other extreme from pure symmetry is pure dependence (sometimes disguised by calling the situation interdependence); but it too is rare. Most cases lie between these two extremes. And that is where the heart of the political bargaining process of interdependence lies.

Power and Interdependence

Power has always been an elusive concept for statesmen and analysts of international politics; now it is even more slippery. The traditional view was that military power dominated other forms, and that states with the most military power controlled world affairs.

But the resources that produce power capabilities have become more complex . . . Hans Morgenthau, author of the leading realist text on international politics, went so far in his reaction to the events of the early 1970s as to announce an historically unprecedented severing of the functional relationship between political, military, and economic power shown in the possession by militarily weak countries of "monopolistic or quasi-monopolistic control of raw materials essential to the operation of advanced economies."

Power can be thought of as the ability of an actor to get others to do something they otherwise would not do (and at an acceptable cost to the actor). Power can also be conceived in terms of control over outcomes. In either case, measurement is not simple. We can look at the initial power resources that give an actor a potential ability; or we can look at that actor's actual influence over patterns of outcomes. When we say that asymmetrical interdependence can be a source of power we are thinking of power as control over resources, or the *potential* to affect outcomes. A less dependent actor in a relationship often has a significant political resource, because changes in the relationship (which the actor may be able to initiate or threaten) will be less costly to that actor than to its partners. This advantage does not guarantee, however, that the political resources provided by favorable asymmetries in interdependence will lead to similar patterns of control over outcomes. There is rarely a one-to-one relationship between power measured by any type of resources and power measured by effects on outcomes. Political bargaining is the usual means of translating potential into effects, and a lot is often lost in the translation.

To understand the role of power in interdependence, we must distinguish between two dimensions, *sensitivity* and *vulnerability*. Sensitivity involves degrees of responsiveness within a policy framework—how quickly do changes in one country bring costly changes in another, and how great are the costly effects? It is measured not merely by the volume of flows across borders but also by the costly effects of changes in transactions on the societies or governments. Sensitivity interdependence is created by interactions within a framework of policies. Sensitivity assumes that the framework remains unchanged. The fact that a set of policies remains constant may reflect the difficulty in formulating new policies within a short time, or it may reflect a commitment to a certain pattern of domestic and international rules.

An example of sensitivity dependence is the way the United States, Japan, and Western Europe were affected by increased oil prices in 1971 and again in 1973–74 and 1975. In the absence of new policies, which could take many years or decades to implement, the sensitivity of these economies was a function of the greater costs of foreign oil and the proportion of petroleum they imported. The United States was less sensitive than Japan to petroleum price rises, because a smaller proportion of its petroleum requirements was accounted for by imports, but as rapid price increases and long lines at gasoline stations showed, the United States was indeed sensitive to the outside change. . . .

Sensitivity interdependence can be social or political as well as economic. For example, there are social "contagion effects," such as . . . the way in which the development of radical student movements during the late 1960s was reinforced by knowledge of each other's activities. The rapid growth of transnational communications has enhanced such sensitivity. Television, by vividly presenting starvation in South Asia to Europeans and Americans about to sit down to their dinners, is almost certain to increase attention to and concern about the issue in European and American societies. Sensitivity to such an issue may be reflected in demonstrations or other political action, even if no action is taken to alleviate the distress (and no economic sensitivity thereby results).

Using the word *interdependence,* however, to refer only to sensitivity obscures some of the most important political aspects of mutual dependence. We must also consider what the situation would be if the framework of policies could be changed. If more alternatives were available, and new and very different policies were possible, what would be the costs of adjusting to the outside change? In petroleum, for instance, what matters is not only the proportion of one's needs that is imported, but the alternatives to imported energy and the costs of pursuing those alternatives. Two countries, each importing 35 percent of their petroleum needs, may seem equally sensitive to price rises; but if one could shift to domestic sources at moderate cost, and the other had no such alternative, the second state would be more *vulnerable* than the first. The vulnerability dimension of interdependence rests on the relative availability and costliness of the alternative that various actors face.

Under the Bretton Woods monetary regime during the late 1960s, both the United States and Great Britain were sensitive to decisions by foreign speculators or central banks to shift assets out of dollars or sterling, respectively. But the United States was less vulnerable than Britain because it had the option (which it exercised in August 1971) of changing the rules of the system at what it considered tolerable costs. The underlying capabilities of the United States reduced its vulnerability, and therefore made its sensitivity less serious politically.

In terms of the costs of dependence, sensitivity means liability to costly effects imposed from outside before policies are altered to try to change the situation. Vulnerability can be defined as an actor's liability to suffer costs imposed by external events even after policies have been altered. Since it is usually difficult to change policies quickly, immediate effects of external changes generally reflect sensitivity dependence. Vulnerability dependence can be measured only by the costliness of making effective adjustments to a changed environment over a period of time.

Vulnerability is particularly important for understanding the political structure of interdependence relationships. In a sense, it focuses on which actors are "the definers of the *ceteris paribus* clause," or can set the rules of the game. Vulnerability is clearly more relevant than sensitivity, for example, in analyzing the politics of raw materials such as the supposed transformation of power after 1973. All too often, a high percentage of imports of a material is taken as an index of vulnerability, when by itself it merely suggests that sensitivity may be high. The key question for determining vulnerability is how effectively altered policies could bring into being sufficient quantities of this, or a comparable, raw material, and at what cost. The fact that the United States imports approximately 85 percent of its bauxite supply does not indicate American vulnerability to actions by bauxite exporters, until we know what it would cost (in time as well as money) to obtain substitutes.

How does this distinction help us understand the relationship between interdependence and power? Clearly, it indicates that sensitivity interdependence will be less important than vulnerability interdependence in providing power resources to actors. If one actor can reduce its costs by altering its policy, either domestically or internationally, the sensitivity patterns will not be a good guide to power resources.

Consider trade in agricultural products between the United States and the Soviet Union from 1972 to 1975. Initially, the American economy was highly sensitive to Soviet grain purchases: prices of grain rose dramatically in the United States. The Soviet Union was also sensitive to the availability of surplus American stocks, since its absence could have internal political as well as economic implications. The vulnerability asymmetries, however, ran strongly in favor of the United States, since its alternatives to selling grain to the USSR (such as government storage, lower domestic prices, and more food aid abroad) were more attractive than the basic Soviet alternative to buying grain from the United States (slaughtering livestock and reducing meat consumption). Thus, as long as the United States government could retain coherent control of the policy—that is, as long as interest groups with a stake in expanded trade did not control it—agricultural trade could be used as a tool in political bargaining with the Soviet Union.

Vulnerability interdependence includes the strategic dimension that sensitivity interdependence omits, but this does not mean that sensitivity is politically unimportant. Rapidly rising sensitivity often leads to complaints about interdependence and political efforts to alter it, particularly in countries with pluralistic political systems. Textile and steel workers and manufacturers, oil consumers, and conservatives suspicious of radical movements originating abroad are all likely to demand government policies to protect their interests. Policymakers and policy analysts, however, must examine underlying patterns of vulnerability interdependence when they decide on strategies. What can they do, at what cost? And what can other actors do, at what cost, in response? Although patterns of sensitivity interdependence may explain where the shoe pinches or the wheel squeaks, coherent policy must be based on an analysis of actual and potential vulnerabilities. An attempt to manipulate asymmetrical sensitivity interdependence without regard for underlying patterns of vulnerability is likely to fail.

Manipulating economic or sociopolitical vulnerabilities, however, also bears risks. Strategies of manipulating interdependence are likely to lead to counterstrategies. It must always be kept in mind, furthermore, that military power dominates economic power in the sense that economic means alone are likely to be ineffective against the serious use of military force. Thus, even effective manipulation of asymmetrical interdependence within a nonmilitary area can create risks of military counteraction. When the United States exploited Japanese vulnerability to economic embargo in 1940–41, Japan countered by attacking Pearl Harbor and the Philippines. Yet military actions are usually very costly; and for many types of actions, these costs have risen steeply during the last thirty years.

Table 1 shows the three types of asymmetrical interdependence that we have been discussing. The dominance ranking column indicates that the power resources provided by military interdependence dominate those provided by nonmilitary vulnerability, which in turn dominate those provided by asymmetries in sensitivity. Yet exercising more dominant forms of power brings higher costs. Thus, *relative to cost,* there is no guarantee that military means will be more effective than economic ones to achieve a given purpose. We can expect, however, that as the interests at stake become more important, actors will tend to use power resources that rank higher in both dominance and cost.

A movement from one power resource to a more effective, but more costly, resource, will be most likely where there is a substantial *incongruity*

Table 1. Asymmetrical Interdependence and its Uses

Source of Interdependence	Dominance ranking	Cost ranking	Contemporary use
Military (costs of using military force)	1	1	Used in extreme situations or against weak foes when costs may be slight.
Nonmilitary vulnerability (costs of pursuing alternative policies)	2	2	Used when normative constraints are low, and international rules are not considered binding (including nonmilitary relations between adversaries, and situations of extremely high conflict between close partners and allies).
Nonmilitary sensitivity (costs of change under existing policies)	3	3	A power resource in the short run or when normative constraints are high and international rules are binding. Limited, since if high costs are imposed, disadvantaged actors may formulate new policies.

between the distribution of power resources on one dimension and those on another. In such a situation, the disadvantaged actor's power position would be improved by raising the level at which the controversy is conducted. For instance, in a concession agreement, a multinational oil company may seem to have a better bargaining position than the host government. The agreement may allow the company to set the level of output, and the price, of the petroleum produced, thus making government revenues sensitive to company decisions. Yet such a situation is inherently unstable, since the government may be stronger on the vulnerability dimension. Once the country has determined that it can afford to alter the agreement unilaterally, it may have the upper hand. Any attempt by the company to take advantage of its superior position on the sensitivity dimension, without recognizing its weakness at the vulnerability level (much less at the level of military force) is then likely to end in disaster.

We conclude that a useful beginning in the political analysis of international interdependence can be made by thinking of asymmetrical interdependencies as sources of power among actors. Such a framework can be applied to relations between transnational actors (such as multinational corporations) and governments as well as interstate relations. Different types of interdependence lead to potential political influence, but under different constraints. Sensitivity interdependence can provide the basis for significant political influence only when the rules and norms in effect can be taken for granted, or when it would be prohibitively costly for dissatisfied states to change their policies quickly. If one set of rules puts an actor in a disadvantageous position, that actor will probably try to change those rules if it can do so at reasonable cost. Thus influence deriving from favorable asymmetries in sensitivity is very limited when the underlying asymmetries in vulnerability are unfavorable. Likewise, if a state chafes at its economic vulnerabilities, it may use military force to attempt to redress that situation as Japan did in 1941; or, it may subtly threaten to use force, as did the United States in 1975, when facing the possibility of future oil boycotts. But in many contemporary situations, the use of force is so costly, and its threat so difficult to make credible, that a military strategy is an act of desperation.

Yet this is not the whole story of power and interdependence. Just as important as understanding the way that manipulation of interdependence can be an instrument of power is an understanding of that instrument's limits. Asymmetrical interdependence by itself cannot explain bargaining outcomes, even in traditional relations among states. As we said earlier, power measured in terms of resources or potential may look different from power measured in terms of influence over outcomes. We must also look at the "translation" in the political bargaining process. One of the most important reasons for this is that the commitment of a weaker state may be much greater than that of its stronger partner. The more dependent actor may be (or appear to be) more willing to suffer. At the politico-military level, the United States' attempt to coerce North Vietnam provides an obvious example.

Yet the point holds even in more cooperative interstate relations. In the Canadian-American relationship, for example, the use or threat of force is virtually excluded from consideration by either side. The fact that Canada has less military strength than the United States is therefore not a major factor in the bargaining process. The Canadians can take advantage of their superior position on such economic issues as oil and natural gas exports without fearing military retaliation or threat by the United States. Moreover, other conditions of contemporary international interdependence tend to limit the abilities of statesmen to manipulate asymmetrical interdependence. In particular, the smaller state may have greater internal political unity than the larger one. Even though the more powerful state may be less dependent in aggregate terms, it may be more fragmented internally and its coherence reduced by conflicts of interest and difficulties of coordination within its own government.

. . . . What we have said is sufficient to indicate that we do not expect a measure of potential power, such as asymmetrical interdependence, to predict perfectly actors' successes or failures at influencing outcomes. It merely provides a first approximation of initial bargaining advantages available to either side. Where predictions based on patterns of asymmetrical interdependence are incorrect, one must look closely for the reasons. They will often be found in the bargaining process that translates power resources into power over outcomes.

International Regime Change

Understanding the concept of interdependence and its relevance to the concept of power is necessary to answering the first major question of this book—what are the characteristics of world politics under conditions of extensive interdependence? Yet as we have indicated, relationships of interdependence often occur within, and may be affected by, networks of rules, norms, and procedures that regularize behavior and control its effects. We refer to the sets of governing arrangements that affect relationships of interdependence as *international regimes*. Although not so obvious as the political bargaining process, equally important to understanding power and interdependence is our second major question: How and why do regimes change?

In world politics rules and procedures are neither so complete nor so well enforced as in well-ordered domestic political systems, and the institutions are neither so powerful nor so autonomous. The weakness of international organizations and the problems of enforcing international law sometimes mislead observers into thinking that international regimes are insignificant, or into ignoring them entirely. Yet although overall global integration is weak, specific international regimes often have important effects on interdependent relationships that involve a few countries, or involve many countries on a specific issue. Since World War II, for instance, specific sets of rules and procedures have been developed to guide states and transnational actors in a wide variety of areas, including aid to less developed countries, environmental protection, fisheries conservation, international food policy, international meteorological coordination, international monetary policy, regulation of multinational corporations, international shipping policy, international telecommunications policy, and international trade. In some cases these regimes have been formal and comprehensive; in others informal and partial. Their effectiveness has varied from issue-area to issue-area and from time to time. On a more selective or regional level, specific groups of countries such as those in the European Community or the Organization for Economic Cooperation and Development (OECD) have developed regimes that affect several aspects of their countries' relationships with each other.

International regimes may be incorporated into interstate agreements or treaties, as were the international monetary arrangements developed at Bretton Woods in 1944, or they may evolve from proposed formal arrangements that were never implemented, as was the General Agreement on Tariffs and Trade (GATT), which derived from the International Trade Organization proposed after World War II. Or they may be merely implicit, as in the postwar Canadian–American relationship. They vary not only in their extensiveness but in the degree of adherence they receive from major actors. When there are no agreed norms and procedures or when the exceptions to the rules are more important than the instances of adherence, there is a *nonregime* situation.

To understand the international regimes that affect patterns of interdependence, one must look . . . at structure and process in international systems, as well as at how they affect each other. The *structure* of a system refers to the distribution of capabilities among similar units. In international political systems the most important units are states, and the relevant capabilities have been regarded as their power resources. There is a long tradition of categorizing the distribution of power in interstate systems according to the number and importance of major actors (for instance, as unipolar, bipolar, multipolar, and dispersed) just as economists describe the structure of market systems as monopolitic, duopolistic, oligopolistic, and competitive. Structure is therefore distinguished from *process,* which refers to allocative or bargaining behavior within a power structure. To use the analogy of a poker game, at the process level analysts are interested in how the players play the hands they have been dealt. At the structural level they are interested in how the cards and chips were distributed as the game started.

International regimes are intermediate factors between the power structure of an international system and the political and economic bargaining that takes place within it. The structure of the system (the distribution of power resources among states) profoundly affects the nature of the regime (the more or less loose set of formal and informal norms, rules, and procedures relevant to the system). The regime, in turn, affects and to some extent governs the political bargaining and daily decision-making that occurs within the system.

Changes in international regimes are very important. In international trade, for example, an international regime including nondiscriminatory trade practices was laid down by the General Agreement on Tariffs and Trade (GATT) in 1947. For almost three decades, the GATT arrangements have constituted a relatively effective international regime. But the last decade, particularly since the first United Nations Conference on Trade and Development in 1964, has been marked by the partly successful efforts of less developed countries to change this regime. More broadly, by the mid-1970s, the demands of less developed countries for a New International Economic Order involved struggles over what international regimes should govern trade in raw materials and manufactures as well as direct foreign investment.

In the two issue areas . . . —money and oceans,—some regime changes have been rapid and dramatic whereas others have been gradual. Dramatic changes took place in international monetary policy in 1914 (suspension of the gold standard); 1931 (abandonment of the gold-exchange standard); 1944 (agreement on the "Bretton Woods System"); and 1971 (abandonment of the convertibility of dollars into gold). Rules governing the uses of the world's oceans changed more slowly, but with significant turning points in 1945 and after 1967. Yet we have no theory in the field of international relations that adequately explains such changes. Indeed, most of our theories do not focus on this question at all.

 . . . . Since world politics varies, over time and from place to place, there is no reason to believe that a single set of conditions will always and everywhere apply, or that any one model is likely to be universally applicable. Thus, before examining the explanatory models, we shall establish the conditions under which they can be expected to apply. . . .

Realism and Complex Interdependence

ROBERT O. KEOHANE AND JOSEPH S. NYE

One's assumptions about world politics profoundly affect what one sees and how one constructs theories to explain events. We believe that the assumptions of political realists, whose theories dominated the postwar period, are often an inadequate basis for analyzing the politics of interdependence. The realist assumptions about world politics can be seen as defining an extreme set of conditions or *ideal type*. One could also imagine very different conditions. In this chapter, we shall construct another ideal type, the opposite of realism. We call it *complex interdependence*. After establishing the differences between realism and complex interdependence, we shall argue that complex interdependence sometimes comes closer to reality than does realism. When it does, traditional explanations of change in international regimes become questionable and the search for new explanatory models becomes more urgent.

For political realists, international politics, like all other politics, is a struggle for power but, unlike domestic politics, a struggle dominated by organized violence. . . . Three assumptions are integral to the realist vision. First, states as coherent units are the dominant actors in world politics. This is a double assumption: states are predominant; and they act as coherent units. Second, realists assume that force is a usable and effective instrument of policy. Other instruments may also be employed, but using or threatening force is the most effective means of wielding power. Third, partly because of their second assumption, realists assume a hierarchy of issues in world politics, headed by questions of military security: the "high politics" of military security dominates the "low politics" of economic and social affairs.

These realist assumptions define an ideal type of world politics. They allow us to imagine a world in which politics is continually characterized by active or potential conflict among states, with the use of force possible at any time. Each state attempts to defend its territory and interests from real or perceived threats. Political integration among states is slight and lasts only as long as it serves the national interests of the most powerful states. Transnational actors either do not exist or are politically unimportant. Only the adept exercise of force or the threat of force permits states to survive, and only while statesmen succeed in adjusting their interests, as in a well-functioning balance of power, is the system stable.

Each of the realist assumptions can be challenged. If we challenge them all simultaneously, we can imagine a world in which actors other than states participate directly in world politics, in which a clear hierarchy of issues does not exist, and in which force is an ineffective instrument of policy. Under these conditions—which we call the characteristics of complex interdependence—one would expect world politics to be very different than under realist conditions.

The Characteristics of Complex Interdependence

Complex interdependence has three main characteristics:

1. *Multiple channels* connect societies, including: informal ties between governmental elites as well as formal foreign office arrangements; informal ties among nongovernmental elites (face-to-face and through telecommunications); and transnational organizations (such as multinational banks or corporations). These channels can be summarized as interstate, transgovernmental, and transnational relations. *Interstate* relations are the normal channels assumed by realists. *Transgovernmental* applies when we relax the realist assumption that states act coherently as units; *transnational* applies when we relax the assumption that states are the only units.

2. The agenda of interstate relationships consists of multiple issues that are not arranged in a clear or consistent hierarchy. This *absence of hierarchy among issues* means, among other things, that military security does not consistently dominate the agenda. Many issues arise from what used to be considered domestic policy, and the distinction between domestic and foreign issues becomes blurred. These issues are considered in several government departments (not just foreign offices), and at several levels. Inadequate policy coordination on these issues involves significant costs. Different issues generate different coalitions, both within governments and across them, and involve different degrees of conflict. Politics does not stop at the waters' edge.

3. Military force is not used by governments toward other governments within the region, or on the issues, when complex interdependence prevails. It may, however, be important in these governments' relations with governments outside that region, or on other issues. Military force could, for instance, be irrelevant to resolving disagreements on economic issues among members of an alliance, yet at the same time be very important for that alliance's political and military relations with a rival bloc. For the former

relationships this condition of complex interdependence would be met; for the latter, it would not.

Traditional theories of international politics implicitly or explicitly deny the accuracy of these three assumptions. Traditionalists are therefore tempted also to deny the relevance of criticisms based on the complex interdependence ideal type. We believe, however, that our three conditions are faily well approximated on some global issues of economic and ecological interdependence and that they come close to characterizing the entire relationship between some countries. . . .

The Political Processes of Complex Interdependence

The three main characteristics of complex interdependence give rise to distinctive political processes, which translate power resources into power as control of outcomes. As we argued earlier, something is usually lost or added in the translation. Under conditions of complex interdependence the translation will be different than under realist conditions, and our predictions about outcomes will need to be adjusted accordingly.

In the realist world, military security will be the dominant goal of states. It will even affect issues that are not directly involved with military power or territorial defense. Nonmilitary problems will not only be subordinated to military ones; they will be studied for their politico-military implications. Balance of payments issues, for instance, will be considered at least as much in the light of their implications for world power generally as for their purely financial ramifications. . . .

In a world of complex interdependence, however, one expects some officials, particularly at lower levels, to emphasize the *variety* of state goals that must be pursued. In the absence of a clear hierarchy of issues, goals will vary by issue, and many not be closely related. Each bureaucracy will pursue its own concerns; and although several agencies may reach compromises on issues that affect them all, they will find that a consistent pattern of policy is difficult to maintain. Moreover, transnational actors will introduce different goals into various groups of issues.

Linkage Strategies

Goals will therefore vary by issue area under complex interdependence, but so will the distribution of power and the typical political processes. Traditional analysis focuses on *the* international system, and leads us to anticipate similar political processes on a variety of issues. Militarily and economically strong states will dominate a variety of organizations and a variety of issues, by linking their own policies on some issues to other states' policies on other issues. By using their overall dominance to prevail on their weak issues, the strongest states will, in the traditional model, ensure a congruence between the overall structure of military and economic power and the pattern of outcomes on any one issue area. Thus world politics can be treated as a seamless web.

Under complex interdependence, such congruence is less likely to occur. As military force is devalued, militarily strong states will find it more difficult to use their overall dominance to control outcomes on issues in which they are weak. And since the distribution of power resources in trade, shipping, or oil, for example, may be quite different, patterns of outcomes and distinctive political processes are likely to vary from one set of issues to another. If force were readily applicable, and military security were the highest foreign policy goal, these variations in the issue structures of power would not matter very much. The linkages drawn from them to military issues would ensure consistent dominance by the overall strongest states. But when military force is largely immobilized, strong states will find that linkage is less effective. They may still attempt such links, but in the absence of a hierarchy of issues, their success will be problematic.

Dominant states may try to secure much the same result by using overall economic power to affect results on other issues. If only economic objectives are at stake, they may succeed: money, after all, is fungible. But economic objectives have political implications, and economic linkage by the strong is limited by domestic, transnational, and transgovernmental actors who resist having their interests traded off. Furthermore, the international actors may be different on different issues, and the international organizations in which negotiations take place are often quite separate. Thus it is difficult, for example, to imagine a militarily or economically strong state linking concessions on monetary policy to reciprocal concessions in oceans policy. . . .

Thus as the utility of force declines, and as issues become more equal in importance, the distribution of power within each issue will become more important. If linkages become less effective on the whole, outcomes of political bargaining will increasingly vary by issue area.

Agenda Setting

Our second assumption of complex interdependence, the lack of clear hierarchy among multiple issues, leads us to expect that the politics of agenda formation and control will become more important. Traditional analyses lead statesmen to focus on politico-military issues and to pay little attention to the broader politics of agenda formation. Statesmen assume that the agenda will be set by shifts in the balance of power, actual or anticipated, and by perceived threats to the security of states. Other issues will only be very important when they seem to affect security and military power. In these cases, agendas will be influenced strongly by considerations of the overall balance of power.

Yet, today, some nonmilitary issues are emphasized in interstate relations at one time, whereas others of seemingly equal importance are neglected or quietly handled at a technical level. International monetary politics, problems of commodity terms of trade, oil, food, and multinational corporations have all been important during the last decade; but not all have been high on interstate agendas throughout that period.

Traditional analysts of international politics have paid little attention to agenda formation: to how issues come to receive sustained attention by high officials. The traditional orientation toward military and security affairs implies that the crucial problems of foreign policy are imposed on states by the actions or threats of other states. These are high politics as opposed to the low politics of economic affairs. Yet, as the complexity of actors and issues in world politics increases, the utility of force declines and the line between domestic policy and foreign policy becomes blurred: as the conditions of complex interdependence are more closely approximated, the politics of agenda formation becomes more subtle and differentiated.

Under complex interdependence we can expect the agenda to be affected by the international and domestic problems created by economic growth and increasing sensitivity interdependence. . . . Discontented domestic groups will politicize issues and force more issues once considered domestic onto the interstate agenda. Shifts in the distribution of power resources within sets of issues will also affect agendas. During the early 1970s the increased power of oil-producing governments over the transnational corporations and the consumer countries dramatically altered the policy agenda. Moreover, agendas for one group of issues may change as a result of linkages from other groups in which power resources are changing; for example, the broader agenda of North–South trade issues changed after the OPEC price rises and the oil embargo of 1973–74. Even if capabilities among states do not change, agendas may be affected by shifts in the importance of transnational actors. . . .

Transnational and Transgovernmental Relations

Our third condition of complex interdependence, multiple channels of contact among societies, further blurs the distinction between domestic and international politics. The availability of partners in political coalitions is not necessarily limited by national boundaries as traditional analysis assumes. The nearer a situation is to complex interdependence, the more we expect the outcomes of political bargaining to be affected by transnational relations. Multinational corporations may be significant both as independent actors and as instruments manipulated by governments. The attitudes and policy stands of domestic groups are likely to be affected by communications, organized or not, between them and their counterparts abroad.

The multiple channels of contact found in complex interdependence are not limited to nongovernmental actors. Contacts between governmental bureaucracies charged with similar tasks may not only alter their perspectives but lead to transgovernmental coalitions on particular policy questions. To improve their chances of success, government agencies attempt to bring actors from other governments into their own decision-making processes as allies. Agencies of powerful states such as the United States have used such coalitions to penetrate weaker governments in such countries as Turkey and Chile. They have also been used to help agencies of other governments penetrate the United States bureaucracy. . . .

The existence of transgovernmental policy networks leads to a different interpretation of one of the standard propositions about international politics—that states act in their own interest. Under complex interdependence, this conventional wisdom begs two important questions: which self and which interest? A government agency may pursue its own interests under the guise of the national interest; and recurrent interactions can change official perceptions of their interests. . . .

The ambiguity of the national interest raises serious problems for the top political leaders of governments. As bureaucracies contact each other directly across national borders (without going through foreign offices), centralized control becomes more difficult. There is less assurance that the state will be united when dealing with foreign governments or that its components will interpret national interests similarly when negotiating with foreigners. The state may prove to be multifaceted, even schizophrenic. National interests will be defined differently on different issues, at different times, and by different governmental units. States that are better placed to maintain their coherence (because of a centralized political tradition such as France's) will be better able to manipulate uneven interdependence than fragmented states that at first glance seem to have more resources in an issue area.

Role of International Organizations

Finally, the existence of multiple channels leads one to predict a different and significant role for international organizations in world politics. Realists in the tradition of Hans J. Morgenthau have portrayed a world in which states, acting from self-interest, struggle for "power and peace." Security issues are dominant; war threatens. In such a world, one may assume that international institutions will have a minor role, limited by the rare congruence of such interests. International organizations are then clearly peripheral to world politics. But in a world of multiple issues, perfectly linked, in which coalitions are formed transnationally and transgovernmentally, the potential role of international institutions in political bargaining is greatly increased. In particular, they help set the international agenda, and act as catalysts for coalition-formation and as arenas for political initiatives and linkage by weak states.

Complex interdependence therefore yields different political patterns than does the realist conception of the world (Table 2 summarizes these differences.) Thus, one would expect traditional theories to fail to explain international regime change in situations of complex interdependence. But, for a situation that approximates realist conditions, traditional theories should be appropriate. . . .

Table 2. Political Processes Under Conditions of Realism
and Complex Interdependence

	Realism	Complex Interdependence
Goals of actors	Military security will be the dominant goal.	Goals of states will vary by issue area. Transgovernmental politics will make goals difficult to define. Transnational actors will pursue their own goals.
Instruments of state policy	Military force will be most effective, although economic and other instruments will also be used.	Power resources specific to issue areas will be more relevant. Manipulation of interdependence, international organizations, and transnational actors will be major instruments.
Agenda formation	Potential shifts in the balance of power and security threats will set agenda in high politics and will strongly influence other agendas.	Agenda will be affected by changes in the distribution of power resources within issue areas; the status of international regimes; changes in the importance of transnational actors; linkages from other issues and politicization as a result of rising sensitivity interdependence.
Linkages of issues	Linkages will reduce differences in outcomes among issue areas and reinforce international hierarchy.	Linkages by strong states will be more difficult to make since force will be ineffective. Linkages by weak states through international organizations will erode rather than reinforce hierarchy.
Roles of international organizations	Roles are minor, limited by state power and the importance of military force.	Organizations will set agendas, induce coalition-formation, and act as arenas for political action by weak states. Ability to choose the organizational forum for an issue and to mobilize votes will be an important political resource.

Revisions of Marxism:
Dependency and World-Systems Theory

Marxism after Marx is replete with lively disagreements and creative tension. Debate has raged among Marxists on a number of key issues: methodology, the nature of the state, the prospects for revolution, to name but a few. Among the various Marxist innovations relevant to IPE, perhaps the most widely discussed are dependency theory and its progeny, world-systems analysis. Although not all dependency theorists would refer to themselves as Marxists, this perspective has been greatly influenced by Marxist thought. Even though world-systems analysis has been criticized as more reminiscent of Adam Smith than Karl Marx, this school of thought has also drawn heavily on the insights of Marxism. Both world-systems analysis and dependency theory grow out of a reconsideration of Marx's expectation that the internationalization of capital would ultimately break down old modes of production and spawn capitalist development around the world.

Theotonio Dos Santos, as well as other early dependency theorists, revises Marx's conclusions on the internationalization of capital. Development in the poorer countries will not follow a course similar to that of Britain and other advanced capitalist countries. Rather, international capital creates a "structure of dependence" and produces: "a situation in which the economy of certain countries is conditioned by the development and expansion of another economy to which the former is subjected." This conditioning effects a transfer of surplus from poor to rich, undermining the accumulation of captial in Third World countries. A typical capitalist mode of production, therefore, cannot develop in many parts of the world. Dependence may take a variety of forms, based upon specific historical circumstances. Colonialism is an obvious and direct type of dependence, but multinational corporations are also a chanism for draining surplus from host countries. Whatever its form, dependence causes underdevelopment and inequality.

The structure of dependence is global in scope. The uneven development inherent in the capitalist mode of production, emphasized by Lenin, is replicated internationally. Thus, less developed countries assume a role akin to an international proletariat in relation to the advanced industrial countries. The world is bifurcated and must be in order for capital to be reproduced. Dos Santos holds that the international structure of dependence renders the Ricardian argument on comparative advantage invalid. World markets cannot promote the interests of the less developed because conditions are rigged in favor of the most developed. Moreover, policy change within the Third World, in and of itself, is unlikely to produce improved economic conditions because: "the productive system in the underdeveloped countries is essentially determined by these international relations." This implies that only fundamental change, and possibly the destruction of exploitative links to the capitalist world economy, will bring genuine development to Asia, Africa, the Middle East, and Latin America.

Dos Santos also points to the political implications of dependence. In some underdeveloped countries, the continuing reliance on agricultural exports, encouraged by international conditions, bolsters the political strength of traditional authoritarian domestic interests. Furthermore, repressive governments are necessary to enforce low wage rates that facilitate the production and export of surplus. As economic prospects for peasants and the working class deteriorate, two political alternatives are possible: increasingly harsh authoritarianism or socialist revolution.

Christopher Chase-Dunn, in explicating the world-systems perspective, takes some exception to dependency theory. From his point of view, greater, though still limited, possibility exists for capital accumulation within certain less developed countries. In addition, socialism is virtually impossible on a national scale; it is not a real alternative to incorporation into the capitalist world-system.

Chase-Dunn comes to these conclusions after a theoretical critique of two interpretations of how politics and economics relate internationally. He argues against "economism," the simplistic assumption that economics determines politics. Conversely, he finds fault with "historicism," which denies the possibility of underlying social laws and is a failing of those who "focus on exclusively political relations." Chase-Dunn contends that economics and politics, while in some ways discrete processes, are interactive and linked by one common logic: the accumulation of capital. Thus, he constructs a "holistic" IPE theory that views an economic process, capital accumulation, as a crucial integrating force. Immanuel Wallerstein's discussion of capitalism as a world-system, is, for Chase-Dunn, the most promising framework for IPE theory, precisely because it avoids the pitfalls of both economism and historicism.

Chase-Dunn states that Wallerstein reconceptualizes capitalism as a mode of production. Where Dos Santos's structure of dependence is based largely upon global economic forces, Wallerstein's capitalist world-system is composed of two complementary subsystems: a world economy driven by the exigencies of accumulation, and an "interstate system" based upon the distribution of politico-military power among states in a formally anarchic environment. Chase-Dunn sees the two as "part of the same interactive socioeconomic system." How are they related to one another? The interstate system provides the political context for a world market. A balance of power among a number of leading states inhibits the development of a single overarching political authority that could subvert international markets. In short, the economic vigor of capitalism depends upon a competitive interstate system.

The interstate system, in turn, is critically influenced by world economic forces. The process of accumulation is responsible for the longevity of the international political system, elements of which can be traced back to early Renaissance Italy. Chase-Dunn argues that unequal economic development favors some countries at some times (i.e., hegemonic core powers), but such advantages are ephemeral in the long term, because technology diffuses globally and offers new economic opportunities to other countries. This economic dynamic prevents any one state from growing into a permanent world-imperial power, thus promoting the maintenance of a competitive interstate system. In addition, contradictions among internationalist and nationalist fractions of the bourgeoisie within a hegemonic core power undermine that state's rise to permanent world domination. Economic forces also work against attempts by second rank powers to impose a new international political order. The Hapsburgs, Louis IV, Napoleon, and Hitler were all unable to harness the changing fortunes of uneven development and failed in their bids for world empire.

World-systems analysis has a broader scope than dependency theory. The latter tends to focus on conditions within underdeveloped countries and correct earlier Marxist predictions regarding the consequences of internationalized capital. Chase-Dunn and Wallerstein, by contrast, focus more on the world-system as a whole and offer insights into a wider range of phenomena: international war among advanced industrial countries as well as development prospects for the Third World.

The Structure of Dependence

THEOTONIO DOS SANTOS

This paper attempts to demonstrate that the dependence of Latin American countries on other countries cannot be overcome without a qualitative change in their internal structures and external relations. We shall attempt to show that the relations of dependence to which these countries are subjected conform to a type of international and internal structure which leads them to under-development or more precisely to a dependent structure that deepens and aggravates the fundamental problems of their peoples.

I. What is Dependence?

By dependence we mean a situation in which the economy of certain countries is conditioned by the development and expansion of another economy to which the former is subjected. The relation of interdependence between two or more economies, and between these and world trade, assumes the form of dependence when some countries (the dominant ones) can expand and can be self-sustaining, while other countries (the dependent ones) can do this only as a reflection of that expansion, which can have either a positive or a negative effect on their immediate development.

The concept of dependence permits us to see the internal situation of these countries as part of world economy. In the Marxian tradition, the theory of imperialism has been developed as a study of the process of expansion of the

Reprinted from *The American Economic Review,* Vol. 16, 1970, with permission of The American Economic Review. © 1970, The American Economic Review.

imperialist centers and of their world domination. In the epoch of the revolutionary movement of the Third World, we have to develop the theory of laws of internal development in those countries that are the object of such expansion and are governed by them. This theoretical step transcends the theory of development which seeks to explain the situation of the underdeveloped countries as a product of their slowness or failure to adopt the patterns of efficiency characteristic of developed countries (or to "modernize" or "develop" themselves). Although capitalist development theory admits the existence of an "external" dependence, it is unable to perceive underdevelopment in the way our present theory perceives it, as a consequence and part of the process of the world expansion of capitalism—a part that is necessary to and integrally linked with it.

In analyzing the process of constituting a world economy that integrates the so-called "national economies" in a world market of commodities, capital, and even of labor power, we see that the relations produced by this market are unequal and combined—unequal because development of parts of the system occurs at the expense of other parts. Trade relations are based on monopolistic control of the market, which leads to the transfer of surplus generated in the dependent countries to the dominant countries; financial relations are, from the viewpoint of the dominant powers, based on loans and the export of capital, which permit them to receive interest and profits; thus increasing their domestic surplus and strengthening their control over the economies of the other countries. For the dependent countries these relations represent an export of profits and interest which carries off part of the surplus generated domestically and leads to a loss of control over their productive resources. In order to permit these disadvantageous relations, the dependent countries must generate large surpluses, not in such a way as to create higher levels of technology but rather superexploited manpower. The result is to limit the development of their internal market and their technical and cultural capacity, as well as the moral and physical health of their people. We call this combined development because it is the combination of these inequalities and the transfer of resources from the most backward and dependent sectors to the most advanced and dominant ones which explains the inequality, deepens it, and transforms it into a necessary and structural element of the world economy.

II. Historic Forms of Dependence

Historic forms of dependence are conditioned by: (1) the basic forms of this world economy which has its own laws of development; (2) the type of economic relations dominant in the capitalist centers and the ways in which the latter expand outward; and (3) the types of economic relations existing inside the peripheral countries which are incorporated into the situation of dependence within the network of international economic relations generated by capitalist expansion. It is not within the purview of this paper to study these forms in detail but only to distinguish broad characteristics of development.

Drawing on an earlier study, we may distinguish: (1) Colonial dependence, trade export in nature, in which commercial and financial capital in alliance with the colonialist state dominated the economic relations of the Europeans and the colonies, by means of a trade monopoly complemented by a colonial monopoly of land, mines and manpower (serf or slave) in the colonized countries. (2) Financial-industrial dependence which consolidated itself at the end of the nineteenth century, characterized by the domination of big capital in the hegemonic centers, and its expansion abroad through investment in the production of raw materials and agricultural products for consumption in the hegemonic centers. A productive structure grew up in the dependent countries devoted to the export of these products . . . producing what ECLA has called "foreign-oriented development" (*desarrollo hacia afuera*). (3) In the postwar period a new type of dependence has been consolidated, based on multinational corporations which began to invest in industries geared to the internal market of underdeveloped countries. This form of dependence is basically technological-industrial dependence.

Each of these forms of dependence corresponds to a situation which conditioned not only the international relations of these countries but also their internal structures: the orientation of production, the forms of capital accumulation, the reproduction of the economy, and, simultaneously, their social and political structure.

III. The Export Economics

In forms (1) and (2) of dependence, production is geared to those products destined for export (gold, silver, and tropical products in the colonial epoch; raw materials and agricultural products in the epoch of industrial-financial dependence); i.e., production is determined by demand from the hegemonic centers. The internal productive structure is characterized by rigid specialization and monoculture in entire regions (the Caribbean, the Brazilian Northeast, etc.). Alongside these export sectors there grew up certain complementary economic activities (cattle-raising and some manufacturing for example) which were dependent, in general, on the export sector to which they sell their products. There was a third, subsistence economy which provided manpower for the export sector under favorable conditions and toward which excess population shifted during periods unfavorable to international trade.

Under these conditions, the existing internal market was restricted by four factors: (1) Most of the national income was derived from export, which was used to purchase the inputs required by export production (slaves, for example) or luxury goods consumed by the hacienda- and mine-owners, and by the more prosperous employees. (2) The available manpower was subject to very arduous forms of superexploitation, which limited its consumption. (3) Part of the consumption of these workers was provided by the subsistence economy, which served as a complement of their income and as a refuge during periods of depression. (4) A fourth factor was to be found in those countries in which land

and mines were in the hands of foreigners (cases of an enclave economy): a great part of the accumulated surplus was destined to be sent abroad in the form of profits, limiting not only internal consumption but also possibilities of re-investment. In the case of enclave economies the relations of the foreign companies with the hegemonic centers were even more exploitative and were complemented by the face that purchases by the enclave were made directly abroad.

IV. The New Dependence

The new form of dependence, (3) above, is in process of developing and is conditioned by the exigencies of the international commodity and capital markets. The possibility of generating new investments depends on the existence of financial resources in foreign currency for the purchase of machinery and processed raw materials not produced domestically. Such purchases are subject to two limitations: the limit of resources generated by the export sector (reflected in the balance of payments, which includes not only trade but also service relations); and the limitations of monopoly on patents which leads monopolistic firms to prefer to transfer their machines in the form of capital rather than as commodities for sale. It is necessary to analyze these relations of dependence if we are to understand the fundamental structural limits they place on the development of these economies.

1. Industrial development is dependent on an export sector for the foreign currency to buy the products utilized by the industrial sector. The first consequence of this dependence is the need to preserve the traditional export sector, which limits economically the development of the internal market by the conservation of backward relations of production and signifies, politically, the maintenance of power by traditional decadent oligarchies. In the countries where these sectors are controlled by foreign capital, it signifies the remittance abroad of high profits, and political dependence on those interests. Only in rare instances does foreign capital not control at least the marketing of these products. In response to these limitations, dependent countries in the 1930's and 1940's developed a policy of exchange restrictions and taxes on the national and foreign export sector; today they tend toward the gradual nationalization of production and toward the imposition of certain timid limitations on foreign control of the marketing of exported products. Furthermore, they seek, still somewhat timidly, to obtain better terms for the sale of their products. In recent decades, they have created mechanisms for international price agreements, and today UNCTAD and ECLA press to obtain more favorable tariff conditions for these products on the part of the hegemonic centers. It is important to point out that the industrial development of these countries is dependent on the situation of the export sector, the continued existence of which they are obliged to accept.

2. Industrial development is, then, strongly conditioned by fluctuations in the balance of payments. This leads toward deficit due to the relations of dependence themselves. The causes of the deficit are three:

a) Trade relations take place in a highly monopolized international market, which tends to lower the price of raw materials and to raise the prices of industrial products, particularly inputs. In the second place, there is a tendency in modern technology to replace various primary products with synthetic raw materials. Consequently the balance of trade in these countries tends to be less favorable (even though they show a general surplus). The overall Latin American balance of trade from 1946 to 1968 shows a surplus for each of those years. The same thing happens in almost every underdeveloped country. However, the losses due to deterioration of the terms of trade (on the basis of data from ECLA and the International Monetary Fund), excluding Cuba, were $26,383 million for the 1951–66 period, taking 1950 prices as a base. If Cuba and Venezuela are excluded, the total is $15,925 million.

b) For the reasons already given, foreign capital retains control over the most dynamic sectors of the economy and repatriates a high volume of profit; consequently, capital accounts are highly unfavorable to dependent countries. The data show that the amount of capital leaving the country is much greater than the amount entering; this produces an enslaving deficit in capital accounts. To this must be added the deficit in certain services which are virtually under total foreign control—such as freight transport, royalty payments, technical aid, etc. Consequently, an important deficit is produced in the total balance of payments; thus limiting the possibility of importation of inputs for industrialization.

c) The result is that "foreign financing" becomes necessary, in two forms: to cover the existing deficit, and to "finance" development by means of loans for the stimulation of investments and to "supply" an internal economic surplus which was decapitalized to a large extent by the remittance of part of the surplus generated domestically and sent abroad as profits.

Foreign capital and foreign "aid" thus fill up the holes that they themselves created. The real value of this aid, however, is doubtful. If over-charges resulting from the restrictive terms of the aid are subtracted from the total amount of the grants, the average net flow, according to calculations of the Inter-American Economic and Social Council, is approximately 54 percent of the gross flow.

If we take account of certain further facts—that a high proportion of aid is paid in local currencies, that Latin American countries make contributions to international financial institutions, and that credits are often "tied"—we find a "real component of foreign aid" of 42.2 percent on a very favorable hypothesis and of 38.3 percent on a more realistic one. The gravity of the situation becomes even clearer if we consider that these credits are used in large part to finance North American investments, to subsidize foreign imports which compete with national products, to introduce technology not adapted to the needs of underdeveloped countries, and to invest in low-priority sectors of the national economies. The hard truth is that the underdeveloped countries have to pay for all of the "aid" they receive. This situation is generating an enormous protest movement by Latin American governments seeking at least partial relief from such negative relations.

3. Finally, industrial development is strongly conditioned by the technological monopoly exercised by imperialist centers. We have seen that the underdeveloped countries depend on the importation of machinery and raw materials for the development of their industries. However, these goods are not freely available in the international market; they are patented and usually belong in the big companies. The big companies do not sell machinery and processed raw materials as simple merchandise: they demand either the payment of royalties, etc., for their utilization or, in most cases, they convert these goods into capital and introduce them in the form of their own investments. This is how machinery which is replaced in the hegemonic centers by more advanced technology is sent to dependent countries as capital for the installation of affiliates. Let us pause and examine these relations, in order to understand their oppressive and exploitative character.

The dependent countries do not have sufficient foreign currency, for the reasons given. Local businessmen have financing difficulties, and they must pay for the utilization of certain patented techniques. These factors oblige the national bourgeois governments to facilitate the entry of foreign capital in order to supply the restricted national market, which is strongly protected by high tariffs in order to promote industrialization. Thus, foreign capital enters with all the advantages: in many cases, it is given exemption from exchange controls for the importation of machinery; financing of sites for installation of industries is provided; government financing agencies facilitate industrialization; loans are available from foreign and domestic banks, which prefer such clients; foreign aid often subsidizes such investments and finances complementary public investments; after installation, high profits obtained in such favorable circumstances can be reinvested freely. Thus it is not surprising that the data of the U.S. Department of Commerce reveal that the percentage of capital brought in from abroad by these companies is but a part of the total amount of invested capital. These data show that in the period from 1946 to 1967 the new entries of capital into Latin America for direct investment amounted to $5,415 million, while the sum of reinvested profits was $4,424 million. On the other hand, the transfers of profits from Latin America to the United States amounted to $14,775 million. If we estimate total profits as approximately equal to transfers plus reinvestment we have the sum of $18,983 million. In spite of enormous transfers of profits to the United States, the book value of the United States direct investment in Latin America went from $3,045 million in 1946 to $10,213 million in 1967. From these data it is clear that: (1) Of the new investments made by U.S. companies in Latin America for the period 1946–67, 55 percent corresponds to new entries of capital and 45 percent to reinvestment of profits; in recent years, the trend is more marked, with reinvestment between 1960 and 1966 representing more than 60 percent of new investments. (2) Remittances remained at about 10 percent of book value throughout the period. (3) The ratio of remitted capital to new flow is around 2.7 for the period 1946–67; that is, for each dollar that enters $2.70 leaves. In the 1960's this ratio roughly doubled, and in some years was considerably higher. The *Survey of Current Business* data on sources and uses of funds for direct North American investment in Latin America in the period 1957–64 show that, of the total sources of direct investment in Latin America, only 11.8 percent came from the United States. The remainder is in large part, the result of the activities of North American firms in Latin America (46.4 percent net income, 27.7 percent under the heading of depreciation), and from "sources located abroad" (14.1 percent). It is significant that the funds obtained abroad that are external to the companies are greater than the funds originating in the United States.

V. Effects on the Productive Structure

It is easy to grasp, even if only superficially, the effects that this dependent structure has on the productive system itself in these countries and the role of this structure in determining a specified type of development, characterized by its dependent nature.

The productive system in the underdeveloped countries is essentially determined by these international relations. In the first place, the need to conserve the agrarian or mining export structure generates a combination between more advanced economic centers that extract surplus value from the more backward sectors, and also between internal "metropolitan" centers and internal inter-dependent "colonial" centers. The unequal and combined character of capitalist development at the international level is reproduced internally in an acute form. In the second place the industrial and technological structure responds more closely to the interests of the multinational corporations than to internal developmental needs (conceived of not only in terms of the overall interests of the population, but also from the point of view of the interests of a national capitalist development). In the third place, the same technological and economic-financial concentration of the hegemonic economies is transferred without substantial alteration to very different economies and societies, giving rise to a highly unequal productive structure, a high concentration of incomes, underutilization of installed capacity, intensive exploitation of existing markets concentrated in large cities, etc.

The accumulation of capital in such circumstances assumes its own characteristics. In the first place, it is characterized by profound differences among domestic wage-levels, in the context of a local cheap labor market, combined with a capital-intensive technology. The result, from the point of view of relative surplus value, is a high rate of exploitation of labour power. . . .

This exploitation is further aggravated by the high prices of industrial products enforced by protectionism, exemptions and subsidies given by the national governments, and "aid" from hegemonic centers. Furthermore, since dependent accumulation is necessarily tied into the international economy, it is profoundly conditioned by the unequal and combined character of international capitalist economic relations, by the technological and financial control of the imperialist centers, by the realities of the balance of payments, by the economic policies of the state, etc. The role of the state in the growth of national and foreign capital merits a much fuller analysis than can be made here.

Using the analysis offered here as a point of departure, it is possible to understand the limits that this productive system imposes on the growth of the internal markets of these countries. The survival of traditional relations in the countryside is a serious limitation on the size of the market, since industrialization does not offer hopeful prospects. The productive structure created by dependent industrialization limits the growth of the internal market.

First, it subjects the labor force to highly exploitative relations which limit its purchasing power. Second, in adopting a technology of intensive capital use, it creates very few jobs in comparison with population growth, and limits the generation of new sources of income. These two limitations affect the growth of the consumer goods market. Third, the remittance abroad of profits carries away part of the economic surplus generated within the country. In all these ways limits are put on the possible creation of basic national industries which could provide a market for the capital goods this surplus would make possible if it were not remitted abroad.

From this cursory analysis we see that the alleged backwardness of these economies is not due to a lack of integration with capitalism but that, on the contrary, the most powerful obstacles to their full development come from the way in which they are joined to this international system and its laws of development.

VI. Some Conclusions: Dependent Reproduction

In order to understand the system of dependent reproduction and the socio-economic institutions created by it, we must see it as part of a system of world economic relations based on monopolistic control of large-scale capital, on control of certain economic and financial centers over others, on a monopoly of a complex technology that leads to unequal and combined development at a national and international level. Attempts to analyze backwardness as a failure to assimilate more advanced models of production or to modernize are nothing more than ideology disguised as science. The same is true of the attempts to analyze this international economy in terms of relations among elements in free competition, such as the theory of comparative costs which seeks to justify the inequalities of the world economic system and to conceal the relations of exploitation on which it is based.

In reality we can understand what is happening in the underdeveloped countries only when we see that they develop within the framework of a process of dependent production and reproduction. This system is a dependent one because it reproduces a productive system whose development is limited by those world relations which necessarily lead to the development of only certain economic sectors, to trade under unequal conditions, to domestic competition with international capital under unequal conditions, to the imposition of relations of superexploitation of the domestic labor force with a view to dividing the economic surplus thus generated between internal and external forces of domination. . . .

In reproducing such a productive system and such international relations, the development of dependent capitalism reproduces the factors that prevent it from reaching a nationally and internationally advantageous situation; and it thus reproduces backwardness, misery, and social marginalization within its borders. The development that it produces benefits very narrow sectors, encounters unyielding domestic obstacles to its continued economic growth (with respect to both internal and foreign markets), and leads to the progressive accumulation of balance-of-payments deficits, which in turn generate more dependence and more superexploitation.

The political measures proposed by the developmentalists of ECLA, UNCTAD, BID, etc., do not appear to permit destruction of these terrible chains imposed by dependent development. We have examined the alternative forms of development presented for Latin America and the dependent countries under such conditions elsewhere. Everything now indicates that what can be expected is a long process of sharp political and military confrontations and of profound social radicalization which will lead these countries to a dilemma: governments of force which open the way to fascism, or popular revolutionary governments, which open the way to socialism. Intermediate solutions have proved to be, in such a contradictory reality, empty and utopian.

Interstate System and Capitalist World-Economy: One Logic or Two?

CHRISTOPHER CHASE-DUNN

This article attempts to clarify a number of issues arising from the critique of Immanuel Wallerstein's (1979) world-system perspective, which is a reinterpretation of the theory of capitalist development inspired by dependency theory. In this perspective, capitalism as a mode of production has always been "imperialistic" in the sense that it constitutes a hierarchical division of labor between core areas and peripheral areas. Wallerstein (1974) has traced the emergence of this capitalist world economy from European feudalism in the long sixteenth century. Viewed as a whole, the modern world-system has exhibited cyclical patterns and secular trends as it has expanded to include the entire globe. . . .

The elaboration of this perspective has raised anew the issue of the relationship between economic and political processes in the capitalist mode of production, and this has occurred in the context of a renaissance of Marxist theories of the state. This burgeoning Marxist literature has, as a response to the so-called "economism" of the Stalinist Third International, emphasized the autonomy of political processes and the "relative autonomy of the state." Though most of this literature has focused on the capitalist state and its relationship to the power of social classes, this emphasis on the autonomy of political factors has spilled over to the more recent critiques of Wallerstein's alleged "economism." These critiques advance claims about the relative autonomy of the interstate system and the processes of geopolitics.

In addition to the neo-Marxist argument a number of non-Marxist political scientists have made similar criticisms. All these authors claim that Wallerstein has reduced the operation of the state system (or "international" system) to a consequence of the process of capital accumulation. Indeed, some have contended that geopolitics and state building are themselves the main motors of the modern world-system. Here I will argue that the capitalist mode of production exhibits a single logic in which both political–military power and the appropriation of surplus value through production and sale on the world market play an integrated role. This article will raise a number of methodological and metatheoretical issues, argue for a redefinition of the capitalist mode of production, and discuss evidence for the interdependence of the interstate system and the capital accumulation process.

Metatheoretical and Methodological Issues

In this article, rather than argue at the philosophical level about economics, politics, and political economy in general, I shall ground the discussion in the particular processes which have been operating in the capitalist world economy since the sixteenth century. However, before I advance my arguments that the state system and the capital accumulation process are part of the same interactive socioeconomic system, I should like to discuss briefly a number of methodological problems raised by this issue.

Why have many theorists who focus on politics tended to adopt a narrowly historicist approach to capitalist development? Marx made a broad distinction between the growth of the forces of production (technology), which occurs in the capital accumulation process, and the reorganization of social relations of production (class relations, forms of property, and other institutions which structure exploitation and the accumulation process). The widening of the world market and the deepening of commodity production to more and more spheres of life have occurred in a series of semiperiodic waves which Kondratieff called "longwaves." These and other economic phenomena seem to be associated with "noneconomic" political events such as wars, revolutions, and so on. This has caused some economists to argue that long waves are not really *economic* cycles at all, but are set off by "exogenous" political events.

The causality and interrelationship among wars, revolutions, and long business cycles is not precisely understood, but the accumulation process expands within a certain political framework to the point where that framework is no longer adequate to the scale of world commodity production and distribution. Thus world wars and the rise and fall of hegemonic core powers (Netherlands, Britain, and the United States) can be understood as the violent reorganization of production relations on a world scale, which allows the accumulation process to adjust to its own contradictions and to begin again on a new scale. Political relations among core powers and the colonial empires which are the formal political structure of core-periphery relations are reorganized in a way which allows the increasing internationalization of capitalist production. Wallerstein's observation that capitalism has always been "international" (and transnational) does not contradict the existence of a long-run increase in the proportion of all production decisions and commodity chains which cross state boundaries.

The above discussion does not establish causal priority between accumulation and political reorganization, but it indicates that these are truly interdependent processes. The tendency toward narrow historicism by those who focus on political events may be due to the greater irregularity of politics and the apparently more direct involvement of human collective rationality in political movements. On the other hand, the overemphasis on determinism and mechanical models of development by those who focus exclusively on "economic" processes may be due to the greater regularity of these phenomena and their lawlike market aggregation of many individual wills uncontrolled by rational collective action. Politics seems less systematic and predictable because human freedom is involved, while economic patterns seem more systematic and determined by forces beyond human will.

These perceptions are correct to a considerable extent precisely because capitalism as a system mystifies the social nature of investment decisions by separating the calculation of profit to the enterprise from the calculation of more general social needs. Anticapitalist movements have tried to reintegrate economics and politics in practice, but up to now the scale of the commodity economy has evaded them. Even the "socialist" states have not succeeded in creating a collectively rational and democratic mode of production. The interstate system itself is the fundamental basis of the competitive commodity economy at the system level. Thus the interaction of world market and state system is fundamental to an understanding of capitalist development and its potential transformation into a more collectively rational system. Neither mechanical determinism nor narrow historicism is useful in this project.

Capitalism as a Mode of Production

The critiques of Wallerstein contain implicit assumptions about the nature of capitalism which tend to conceptualize it as an exclusively "economic" process. Skocpol (1979:22) formulates the issue by arguing that Wallersteinians "assume that individual nation-states are instruments used by economically dominant groups to pursue world-market oriented development at home and international economic advantages abroad." She continues, explaining her own position:

> But a different perspective is adopted here, one which holds that nation-states are, more fundamentally, organizations geared to maintain control of home territories and populations and to undertake actual or potential military competition with other states in the international system. The international states system as a transnational structure of military competition was not originally created by capitalism. Throughout modern world history, it represents an analytically autonomous level of transnational reality—*interdependent* in its structure and dynamics with world capitalism, but not reducible to it.

[Other Scholars] argue even more strongly for the autonomy of the state system in opposition to what they see as Wallerstein's economic reductionism. These authors raise the important question about the extent to which it is theoretically valuable to conceptualize economic and political processes as independent sub-systems, but in so doing they oversimplify Wallerstein's perspective. Wallerstein's work suggests a reconceptualization of the capitalist mode of production itself, such that references to capitalism do not point simply to market-oriented strategies for accumulating surplus value. According to Wallerstein, the capitalist mode of production is a system in which groups pursue both political–military and profitable strategies, and the winners are those who effectively combine the two. Thus the state system, state building, and geopolitics are the political side of the capitalist mode of production.

This mode of production is a feature of the whole world-system, not its parts. Wallerstein distinguishes between world empires and world economies. A world empire is a socioeconomic system in which the economic division of labor is incorporated within a single overarching state apparatus. A world economy is an economic division of labor which is overlaid by a multicentric system of states. The political system of capitalism is not the state, but the larger competitive state system. Particular states exhibit more or fewer tendencies toward politicomilitary aggrandizement or free market accumulation depending on their position in the larger system, and the system as a whole goes through periods in which state power is more generally employed versus periods in which a relatively free world market of commodity exchange comes to the fore.

Hegemonic core states with a clear competitive advantage in production advocate free trade. Similarly, peripheral states in the control of peripheral capitalist producers of low-wage goods for export to the core support the "open economy" of free international exchange. Smaller core states heavily dependent on international trade tend to support a liberal economic order. Semi-peripheral states and core states contending for hegemony utilize protectionism and mercantilist monopoly to protect and expand their access to world surplus value. Periods of rapid worldwide economic growth are characterized by a relatively unobstructed world market of commodity exchange as the interests of consumers in low prices come to outweigh the interests of producers in protection. In periods of stagnation, political power is more frequently utilized to protect shares of a diminishing pie.

How does this conceptualization of the capitalist mode of production differ from that of Marx? Marx's model of capital accumulation assumed an institutional basis in which the state played no direct role in production, but maintained the class relations necessary for private accumulation to proceed. His abstract model, explicated in Volume 1 of *Capital,* assumed this "caretaker" state, and also a class structure composed of only two classes, capitalists and proletarians. The world-system perspective, in attempting to come to grips with the realities of capitalist development which have become apparent since Marx's time, seeks to integrate the state and class relations into the accumulation model. Class structures and states are not now seen as merely "historical" forms, the product of impossibly complex processes, but as related in a systematic way to the process of accumulation.

The world-system perspective revises Marx's definition of the capitalist mode of production as follows. The mode of production is thought to be a characteristic of the whole effective division of labor, that is, the world economy. Therefore, the capitalist mode of production includes commodity producers employing wage labor in the core areas and coerced labor in the peripheral areas. Peripheral areas are not seen as "precapitalist" but rather as integrated, exploited, and essential parts of the larger system. Capitalist production relations, in this view, are not limited to wage labor (which is nevertheless understood to be very important to the expanded reproduction of the core areas), but rather are composed of the combination of wage labor with coerced labor in the periphery. This combination is accomplished, not only by the world market exchange of commodities, but also by the forms of political coercion which the core powers often exercise over peripheral areas. The state, and the system of competing states which compose the world polity, constitute the basic structural support for capitalist production relations. Marx saw that the state stood behind the opaque capital/wage-labor relationship in nineteenth-century England. The more direct involvement of the state in the extraction of surplus product with slave labor or serf labor is but a more obvious example of the way in which the state stands behind production relations.

If the state is often directly involved in extraction of surplus product, what is the difference between capitalist production relations and labor extraction in precapitalist agrarian empires? The difference is that the capitalist world economy has no overarching single state which encompasses the entire arena of economic competition. Rather than a world state, there is the "international system" of competing states operating within the world market. Thus state power is used to extract labor power (more directly in the periphery than in the core) but the competitive nature of the state system prevents any single state from maintaining a systemwide monopoly and subjects producers to the necessity of increasing productivity in order to maintain or increase their shares of the world product.

The state system enforces the capital/wage-labor relationship in the core, the coerced labor extraction in the periphery, and the shifting forms of extraction between the core and the periphery, and constitutes the basis of production relations for the capitalist system. States are organizations within the arena of competition which are often utilized by the classes that control them to expropriate shares of the world surplus product. Market forces are either supported or distorted depending on the world market position of the classes controlling a particular state. Thus hegemonic core states and peripheral states (controlled by classes producing raw materials for export to the core) implement, and try to influence others to implement, free trade policies. States in which producers are seeking to protect home markets against cheap foreign goods erect tariff barriers and other political controls on the world market forces. Thus both political organizations and economic producers are subjected to a long-run "competing down" process, whereas in the ancient empires the monopoly of violence held by a single center minimized both market and political competition among different organizational forms.

In the capitalist world economy, state structures themselves are submitted to a political version of the "competing down" process which occurs in the realm of the market. Inefficient state structures, ones that tax their producers too heavily or do not spend their revenues in ways which facilitate politico-economic competition in the world economy, lose the struggle for domination. In Marxist theoretical terms, the state system produces an equalization of surplus profits, the profits which return because political coercion enforces monopolies. There are no complete long-run monopolies. Even the largest organizations (both states and firms) are subjected to the pressures of politico-economic competition.

It has been pointed out that not all precapitalist modes of production were world empires. Indeed, it was European feudalism's "parcellization of sovereignty" which allowed the capitalist mode of production to become dominant. The possibilities for political competition and the space for new departures in social relations which such a decentralized political structure allowed were a fertile context for the emergence of capitalism. Classical European feudalism, although somewhat integrated culturally, and sharing a common political matrix, did not have more than a rudimentary division of labor across the manors and towns. The growth of commodity production, merchant capital and "industrial" capital producing manufactures for the emerging local and long-distance markets transformed the "stateless" classical feudalism into the capitalist world economy and interstate system of the sixteenth century. The dynamic of mercantile competition between both public and private enterprises in the long sixteenth century, together with the emergence of a core-periphery hierarchy, led Wallerstein to argue that the capitalist system was then born. . . .

Most social theorists have correctly identified the differentiation between economic and political institutions as a key feature of capitalism, but, since Adam Smith, this separation has been primarily identified with the separation between private and public spheres within nation-states. The contemporary growth of state capitalism, and Wallerstein's reinterpretation of seventeenth- and eighteenth-century mercantilism, imply that the main locus of the differentiation between economic and political institutions is at the level of the world economy as a whole—that is, the distinction between the state system and the world division of labor. The public-private separation was an important political issue for the triumph of industrial capital over the state-institutionalized interests of older agrarian capital in the eighteenth and nineteenth centuries, especially in the core states. But the more direct role of the state in fostering national economic development in most of the world economy belies the exclusive identification of capitalism with private ownership.

In the competitive state system it has been impossible for any single state to monopolize the entire world market, and to maintain hegemony indefinitely. Hegemonic core powers, such as Britain and the United States, have, in the long run, lost their relative domination to more efficient producers. This means that, unlike the agrarian empires, success in the capitalist world-system is based on a combination of effective state power and competitive advantage in production. The extraction of surplus product is based on two legs: the ability to use political power for the appropriation of surplus product; and the ability to produce efficiently for the competitive world market. This is not the state-centric system which some analysts describe, because states cannot escape, for long, the competitive forces of the world economy. States that attempt to cut themselves off or who overtax their domestic producers condemn themselves to marginality. On the other hand, the system is not simply a free world market of competing producers. The interaction between political power and competitive advantage is a delicate balance.

Skocpol (1977) has argued that Wallerstein is wrong in contending that core states are strong states. She correctly points out that the most successful core states have been those that combined a relatively strong world-economy-oriented bourgeoisie with a relatively decentralized state. . . . This problem can best be handled by observing that there is a certain differentiation among core states in the type of development path followed. Some rely more on geopolitical military advantage and centralized and effective fiscal structures, while others—the more successful ones—rely on low over-head decentralized states which act efficiently to protect and extend the vital business interests of their national capitalists. . . .

Thus, core states are strong relative to peripheral states, but some are stronger vis-à-vis their own internal capitalist class fractions than others. The most successful core nations have achieved their hegemony by having strong and convergent business class interests which unified state policy behind a drive for successful commodity production and trade with the world market. Second-runners have often achieved some centrality in the world economy by relying on a more state-organized attempt to catch up with the "caretaker" states in terms of political and economic hegemony. Skocpol's (1977) emphasis on the less autocratic form of development in the most successful states does not lend support to the contention that geopolitics and capital accumulation are autonomous from one another, although success is perhaps not the best criterion for determining the nature of capitalism as a system. For this we need to focus on the dynamics and relationships operating in the system as a whole, not in its parts, the national societies.

It may be argued that the existence of states which successfully follow an exclusively political–military development path is evidence in favor of the thesis that geopolitical and economic processes operate independently. The existence of such a development path is unquestionable (for example, Prussia, Sweden, Japan, U.S.S.R.), but the upward mobility of these states was certainly conditioned by their context, a world economy in which commodity production and capitalist accumulation were becoming general. If all states had followed such a path, the modern world-system would be a very different kind of entity. It is argued below that the reproduction and expansion of the kind of interstate system which emerged in Europe requires the institutional forms and dynamic processes which are associated with commodity production and capitalist accumulation. First, though, let us discuss the ways in which the interstate system helps to preserve the dynamics of the capitalist process of accumulation.

The Reproduction of Capitalist Accumulation

The competitive state system serves several functions which allow the capitalist accumulation process to overcome temporarily the contradictions it creates, and to expand. The balance of power in the interstate system prevents any single state from controlling the world economy, and from imposing a political monopoly over accumulation. This means that "factors of production" cannot be constrained to the degree that they could be if there were an overarching world state. Capital is subjected to controls by states, but it can still flow from areas where profits are low to areas where profits are higher. This allows capital to escape the political claims which exploited classes attempt to impose on it. If workers are successful in creating organizations which enable them to demand higher wages, or if communities demand that corporations spend more money on pollution controls, capital can usually escape these demands by moving to areas where there is less opposition. This process can also be seen to operate within federal nation-states. Class struggles are most often oriented toward and constrained within particular territorial state structures.

Thus the state system provides the political underpinning of the mobility of capital, and also the institutional basis for the continuing expansion of capitalist development. States which successfully prevent capital from migrating do not necessarily solve this problem, because capital from other sources may take advantage of the less costly production opportunities outside the national boundaries, and push the domestic products out of the world market.

This implies that capitalism is not possible in the context of a single world state, or rather that such a system would eventually develop a political regulation of resource allocation which would more regularly and fully include social desiderata in the calculation of investment decisions. The dynamic of the present system, in which profit criteria and national power are the main controllers of resource use, would be transformed into a system which balances development according to a calculation of the individual and collective use values of human society. This system, which we can call socialism, would not constitute a utopia in which the problem of production has been completely solved, but in political struggles for resources, which would be oriented toward a single overarching world government, would exhibit a very different long-run dynamic of political change and economic development than that which has characterized the capitalist world economy.

Reproduction of the Interstate System

Thus I am arguing that the interstate system is important for the continued viability of the capital accumulation process. But is the accumulation process equally as important for the generation and reproduction of the interstate system? First, what do I mean by reproduction of the interstate system? . . . By an interstate system, I mean a system of unequally powerful and competing states in which no single state is capable of imposing control on all others. These states are in interaction with one another in a set of shifting alliances and wars, and changes in the relative power of states upsets any temporary set of alliances, leading to a restructuring of the balance of power. If such a system disintegrates due to the dissolution of the states, or due to the complete elimination of economic exchanges between the national territories, or as a result of the imposition of domination by a single overarching state, the system can be said to have fundamentally changed (that is, to be transformed, not reproduced). . . .

Skocpol contends that the state system predates the emergence of capitalism and implies that this is evidence of its relative autonomy. It is clearly the case that multistate systems exhibiting some of the tendencies of the European balance of power existed prior to the emergence of the capitalist mode of production. These state systems were unstable, however, and tended either to become world empires or to disintegrate. The multicentric "international system" which developed among the Italian city-states and their trade partners in the East and West invented many of the institutions of diplomacy and shifting alliance which were later adopted by the European states. . . .

Many of the capitalistic financial and legal institutions later elaborated in the European capitalist world economy were invented in the Italian city-states. Wallerstein contends that the Christian Mediterranean constituted a kind of interstitial proto-capitalist world economy. Analogous to Marx's analysis of merchant capitalism, the Mediterranean world economy—though developing the seeds of capitalist production with labor as a commodity—was primarily based on the exchange of preciosities among social systems which were not integrated into a single-commodity economy. Nevertheless this proto-capitalist world economy succeeded in developing several institutional features which were only later fully articulated in the capitalist mode of production that emerged in Europe and Latin America in the sixteenth century. One of these was the state system, which . . . only became stabilized after its emergence in Europe.

Does the continuity of the Italian state system, its failure to develop into a world empire, constitute a case for the independence of the state system? No—its incorporation into the emerging capitalist world economy allowed the Italian system to evade this fate.

Skocpol's contention that nation-states predated the dominance of capitalism is clearly correct. Medieval states were present in precapitalist England and France. However, the emergence of the interstate system is another matter.

The balance-of-power system defined above emerged first among the Italian city-states and later in the Europe of the long sixteenth century, contemporaneous with the emerging dominance of capitalism as a world economy.

One clue to the dependence or independence of the state system is its ability to reproduce itself, or to weather crises without either becoming transformed into a world empire, or experiencing disintegration of its division of labor and network of economic exchange. Wallerstein's analysis of the Habsburg failure to transform the still shaky sixteenth century capitalist world economy into a world empire is asserted to demonstrate the strength of capitalism in reproducing the state system. I will discuss the later points at which such challenges to the state system were mounted (Louis XIV's, the Napoleonic wars, and the twentieth-century world wars) and the causes of continuity of the state system. . . .

The European world-system became the global capitalist system in a series of expanding waves which eventually incorporated all the territories and peoples of the earth. Although political–military alliances with states external to the system occurred after the sixteenth century, they were never again so crucial to the survival and development of the system. But the system continued to face challenges of survival based on its internal contradictions. Uneven economic development and the vast expansion of productive forces outstripped the structures of political power, causing violent reorganizations of the state system (hegemonic wars) to accommodate new levels of economic development. This process can be seen in the cycles of core competition, the rise and fall of hegemonic core states, which have accompanied the expansion and deepening of the capitalist mode of production.

After the failure of the Habsburgs there have been three other attempts to impose a world empire on the capitalist world economy: those of France under Louis XIV and Napoleon, and that of Germany and its allies in the twentieth-century world wars. Each of these came in a period when the hegemonic core power was weak. Louis XIV attempted to expand his monarchy over the whole of the core powers during the decline of Dutch hegemony. Napoleon's attempt came while Britain was still emerging as the hegemonic power. The German attempts came after Britain's decline and before the full emergence of the United States. In these three instances we may see a threat to the state system and to the existence of the capitalist world economy.

One striking thing about these events is that they were not perpetrated by the hegemonic core powers themselves, but by emerging second runners among the competing core states. This raises the question of why hegemonic core powers do not try to impose imperium when it becomes obvious that their competitive advantage in production is waning. Similarly we may ask, why opposing forces were able to prevent the conversion of the system into a single empire. To both of these questions I would answer that the transitional structures associated with the capitalist commodity economy operated to tip the balance in favor of preserving the state system.

Hegemonic core states often use state power to enforce the interests of their "own" producers, although typically they do not rely on it as heavily as other competing core states. However, a hegemonic core power begins to lose its competitive edge in production with the spread of production techniques and the equalization of labor costs which accompany the growth of new core production in other areas. The profit rate differentials change such that capital is exported from the hegemonic core state to areas where profit rates are higher. This reduces the level at which the capitalists of the hegemonic core state will support the "economic nationalism" of their home state. Their interests come to be spread across the core. In other words, hegemonic core states develop fractions of their capitalist classes having different interests. There evolves a fraction of "international capitalists" who support peace and supranational federation, and "national" capitalists who seek protection and politicomilitary expansion. This explains the ambivalent and contradictory policies of hegemonic core powers during the periods of their decline.

Why have the second-running core powers who seek to impose imperium on the world economy failed? Most theorists of the state system have not addressed this question as such. The balance-of-power idea explains why, in a multicentric system, alliances between the most powerful actors weaken. Coalitions in a triad, for example, balance the two weakest actors against the strongest. However, this alliance falls apart when the stronger of the partners gains enough to become the strongest single actor, because the ally can gain more by allying with the declining former power than by sticking to the original alliance. This simple game theory is extended to the state system by the theorists of equilibrium, but it does not answer our question substantively. Again, it is not the most powerful actor that tries to impose imperium, but upwardly mobile second runners with less than their "fair" share of political influence over weaker areas of the globe. . . .

Of course, we are not considering strictly historical explanations which make use of uniquely historical factors: such a theoretical maneuver (or rather an atheoretical maneuver) is easy to accomplish when one is explaining only four "events." Here we seek an explanation of what seems to be a regularity of the system from our understanding of the logic of the system itself.

Morganthau invokes the notion of a normatively organized liberal world culture which successfully mobilizes counterforce against the threat to the balance-of-power system. This conceptualization of the world-system as a normatively integrated system is shared by more recent authors who seek to extend modernization theory to the world-system. Wallerstein's perspective emphasizes that the capitalist world economy is not primarily a normatively integrated system. In Wallerstein's broad typology of social systems, social systems based on normatively regulated reciprocal exchange (termed "mini-systems" by Wallerstein) no longer exist except in vestigial form in the family and as symbolic subsystems of the present world economy, which are not determinant of its developmental tendencies. While Wallerstein does not deny that some normative patterns are generalized across the system, he focuses on the face that culture tends to follow state boundaries so that the system remains primarily multicultural.

The linguistic boundaries of the world culture are formed and reformed primarily by the process of nation building and associated state formation. These processes are somewhat similar across the system, and the national cultures which come to exist have an isomorphic character. These facts do not contradict the main point that normative integration at the system level remains weak, although growing. From this perspective it is farfetched to explain the failure of empire in terms of commitment to shared norms. The ideologies employed by the second runners undoubtedly played some part in their inability to mobilize support, but this is unlikely to have been the most important factor.

I argue that both the attempts and the failures of imperium can be understood as responses to the pressures of uneven development in the world economy, albeit somewhat reactive responses. We have already noted that the attempts were fomented, not by the most powerful states in the system, but by emerging second-running core powers contending for hegemony. One striking aspect of all four cases is their appearance, in retrospect, of wild irrationality in terms of their attempt to use military power to conquer areas much too vast to subdue, let alone to exploit effectively. In this we may see the weakness of the strategy of politicomilitary domination unaccompanied by a strategy of competitive production for sale on the market. The countries that adopted the strategy of aggrandizement reached far beyond their own capacities and failed to generate sufficient support from allied countries. This second condition bears further examination. Why did the French and German imperiums not receive more support? Potential allies, in part, doubted the extent to which their interests would be protected under the new imperium, and the path of capitalist growth in the context of the multicentric system appeared preferable to the emerging bourgeoisies of potential allied states.

If I am correct, the interstate system is dependent on the institutions and opportunities presented by the world market for its survival. There are two main characteristics of the interstate system which need to be sustained: the division of sovereignty in the core (interimperial rivalry) and the maintenance of a network of exchange among the states. The nature of the world economy assures that states will continue to exchange due to natural and socially created comparative advantages in production. Withdrawal from the world market can be accomplished for short periods of time, but it is costly and unstable. Even the "socialist" states which have tried to establish a separate mode of production have eventually returned to production for, and exchange with, larger commodity markets.

The maintenance of interimperial rivalry is facilitated by a number of institutional processes. At any point in time national sentiments, language, and cultural differences make supranational integration difficult, as is well illustrated by the EEC. These "historical" factors may be traced back to the long-run processes of state formation and nation building, and these processes have themselves been conditioned by the emergence of the commodity economy over the past 500 years.

The main institutional feature of the world economy which maintains interimperial rivalry is the uneven nature of capitalist economic development.

As discussed above, hegemonic core powers lose their competitive advantage in production to other areas. This causes the export of capital, which restrains the hegemonic power from attempting to impose political imperium. Second-running challengers, who do try to impose imperium, cannot gain sufficient support from other core allies to win, or at least, historically, they have not been able to do so. This is in part because the potential for further expansion and deepening of the commodity economy, and growth and development in the context of a decentralized interstate system, appears greater to potential allies than the potential for political and economic power within the proposed imperium. It is the very success of capitalist development in the past which preserves the interstate system. Success stories in the uneven development history of the world economy are frequent enough to prevent imperium.

Despite the contentions of many Marxists that the increased size of firms has led to a new stage of "monopoly" capitalism, the conception of capitalism proposed here—which incorporates state capitalism and political control of home markets as normal instances of capitalist competition for shares of the world market—implies that the basic tendencies of the accumulation of capital have been with us since the sixteenth century. This does not deny the increased political density of controls on accumulation which has accompanied the secular increase in the power of states over the process of production. However, this increased density of political control, including the resistance of peripheral states to unbridled exploitation by the core, has not altered the basic nature of the larger system.

The transformation of capitalism cannot be accomplished by the emergence of state control as long as the interstate system remains unregulated by a world government. Proponents of the relative autonomy thesis might point to the bickering and bloody competition among the "socialist" states as evidence of the independence of the interstate system from the capitalist mode of production. My position is that these states remain part of the larger capitalist world-system. This does not mean that the institutional experiments in the "socialist" states have no meaning for the transformation of capitalism, but that they, themselves, do not constitute that transformation.

The current return to mercantilist international economic policy and the rising nationalism and conflict among states can be understood primarily in terms of the repetition of earlier cyclical phases of the world-system. The capitalist mode of production with its logic, including the logic of the interstate system, remains very much the dominant source of developmental tendencies in the contemporary world-system.

The conclusion I wish to reach here is by no means definitive. The usefulness of theorizing about the modern world-system in terms of a single logic versus political and economic subsystems can be decided only when competing theories have been formulated with enough clarity to allow them to be systematically subjected to evidence. My goal has been to present an alternative way of conceptualizing capitalist development in order to stimulate further discussion and research. The attempt to create a reintegrated interdisciplinary science of political economy is a necessary step in the project to understand (and influence) the directions and potentialities of our present collective history.

Bibliography

Skocpol, Theda. "Wallerstein's World Capitalist System. A Theoretical and Historical Critique." *American Journal of Sociology* 82:5 (March 1977): 1075–90.
———. *States and Social Revolutions.* New York: Cambridge University Press, 1979.
Wallerstein, I. *The Modern World System.* New York: Academic Press, 1974, 1980.
———. *The Capitalist World Economy.* Cambridge: Cambridge University Press, 1979.

6

The Return to Statist Theories
of International Political Economy

In the 1970s, disillusionment with liberal and Marxist approaches to IPE led some theorists back to mercantilist themes. Their primary concern was how political action, centering on the definition of state interests and the use of state power, influences international economic relations. These state-oriented arguments discard a number of classic mercantilist ideas, such as the importance of specie, and they reject the normative superiority of economic nationalism. A link to the mercantilist past can be found, however, in the definition of politics as potentially autonomous from class interests and market forces. For the neomercantilist, politics is neither epiphenomenal nor irrational.

Robert Gilpin is instrumental in the revival of state-power analyses of IPE. He situates himself in contrast to Marxist and liberal perspectives, arguing vigorously against "transnational ideologists," who are overly optimistic in their assessment of interdependence, and "revisionist" theories of U.S. foreign policy, which explain strategic actions as extensions of economic interests. Instead, he posits an interactive IPE theory which, in contradistinction to Chase-Dunn's position, emphasizes the importance of political and security interests in shaping economic outcomes. Thus, he sees transnational actors, such as multinational corporations, as dependent upon the political context created by the most powerful states. To make this point, he examines British and American moments of international preeminence.

In his discussion of the United States, Gilpin argues that the Cold War was not simply a result of the expanding power of the U.S. bourgeoisie. He illustrates how, at times, U.S. state action contradicted national capitalist interests, particularly in regard to the trade discrimination U.S. leaders accepted as a cost for the political benefits of an economically integrated Europe. Moreover, he suggests that peculiar strategic circumstances, the reliance on air and sea power as opposed to a large standing army, critically influenced postwar U.S. international economic policy. Security interests of the state, not economic interests of major corporations, motivated U.S. policy toward Europe and Japan.

How do transnational corporations influence international politics? Gilpin holds that the global corporation is an American phenomenon that has, ironically, inspired an even greater role of political regulation in world markets. In response to the American challenge, European states have promoted internationalization of their national corporations and have attempted to limit the power of U.S. firms. This does not mean that economic forces are irrelevant to international relations. Gilpin recognizes the economic and technological dynamics that have given rise to transnational corporations are not wholly determined by political conditions. His primary concern, however, is to illustrate how global economics is constrained by the structure of international politics. In looking to the future, he suggests that changes in the political environment could undermine transnationalism: "the determining consideration will be the diplomatic and strategic interests of the dominant powers."

Peter Katzenstein takes the question of state power in another direction. He, too, is responding to interdependence theory. His analysis accepts two points made by the interdependence school: the difficulty of maintaining the analytic distinction of "international" and "domestic;" and the increasing salience of economic issues versus military affairs. The first point is especially important for Katzenstein. Although he recognizes the significance of hegemonic international power, he contends that the ways in which domestic political structures influence national responses to global conditions has been unduly ignored by IPE analysts. Thus, he shifts the level of analysis away from the international regime focus of interdependence and concentrates on national political forces. Katzenstein also disputes dependency theory's emphasis on North–South relations. While agreeing that the advanced industrial countries dominate the less developed countries, he argues that dependency theory disregards consequential differences among the more powerful countries.

Katzenstein distinguishes his analysis of domestic political structures from the bureaucratic politics approach. The latter is based upon the U.S. experience. Since the structure of the U.S. state is not replicated in other advanced industrial countries, the utility of the bureaucratic politics model is limited. Katzenstein considers two types of domestic structures, ruling coalitions and policy networks, and how these vary from one country to another. A ruling coalition is composed of both state institutions and organized social forces and is responsible for the articulation of policy objectives. Policy networks, again combining elements of both state and society, determine the policy instruments available for implementation. Thus, a country's response to international conditions, as reflected in its policy objectives and policy instruments, is shaped by the structure of its ruling coalition and policy networks. By way of example, Katzenstein suggests that in the United States and the United Kingdom, the ruling coalition of business interests and the state is "relatively unfavorable to state officials and the policy network linking the public with the private sector is relatively fragmented." In Japan, the ruling coalition affords state bureaucrats greater power, and policy networks are more coherent. In short, the contrasting contemporary experiences of the United States and Japan can be explained by domestic political structures.

The Politics of Transnational Economic Relations

ROBERT GILPIN

I

. . . . International society, we are told, is increasingly rent between its economic and its political organization. On the one hand, powerful economic and technical forces are creating a highly integrated transnational economy, blurring the traditional significance of national boundaries. On the other hand, the nation-state continues to command men's loyalties and to be the basic unit of political decision.

. . . . In specific terms the issue is whether the multinational corporation has become or will become an important actor in international affairs, supplanting, at least in part, the nation-state. If the multinational corporation is indeed an increasingly important and independent international actor, what are the factors that have enabled it to break the political monopoly of the nation-state? What is the relationship of these two sets of political actors, and what are the implications of the multinational corporation for international relations? Finally, what about the future? If the contemporary role of the multinational corporation is the result of a peculiar configuration of political and economic factors, can one foresee the continuation of its important role in the future?

Fundamental to these rather specific issues is a more general one raised by the growing contradiction between the economic and political organiza-

Reprinted from *International Organization,* Vol. XXV, No. 3, Summer 1971, Robert Gilpin, "The Politics of Transnational Economic Relations" by permission of The MIT Press, Cambridge, MA.

tion of contemporary international society. This is the relationship between economic and political activities. While the advent of the multinational corporation puts it in a new guise, the issue is an old one. It was, for example, the issue which in the nineteenth century divided classical liberals like John Stuart Mill and the German Historical School represented by Georg Friedrich List. Whereas the former gave primacy to economics and the production of wealth, the latter emphasized the political determination of economic relations. As this issue is central to the contemporary debate on the implications of the multinational corporation for international relations, I would like to discuss it in brief outline.

The classical position was, of course, first set forth by Adam Smith in *The Wealth of Nations*. While Smith appreciated the importance of power, his purpose was to inquire into the nature and causes of wealth. Economic growth, Smith argued, is primarily a function of the extent of the division of labor which in turn is dependent upon the scale of the market. Much of his attack, therefore, was directed at the barriers erected by feudal principalities and mercantilist states against the free exchange of goods and the enlargement of markets. If men are to multiply their wealth, Smith argued, the contradiction between political organization and economic rationality had to be resolved in favor of the latter.

Marxism, the rebellious ideological child of classical liberalism, erected the concept of the contradiction between economic and political relations into a historical law. Whereas classical liberalism held that the requirements of economic rationality *ought* to determine political relations, the Marxist position was that the mode of production *does* determine the superstructure of political relations. History can be understood as the product of the dialectical process—the contradiction between evolving economic forces and the sociopolitical system.

Although Karl Marx and Friedrich Engels wrote amazingly little on the subject of international economies, Engels in his famous polemic, *Anti-Dühring,* dealt explicitly with the question of whether economics or politics was primary in determining the structure of international relations. Karl Dühring's anti-Marxist theory maintained that property relations resulted less from the economic logic of capitalism than from extraeconomic political factors. Engels, on the other hand, using the example of the unification of Germany in his attack on Dühring, argued that economic factors were primary.

Engels argued that when contradictions arise between economic and political structures, political power adapts itself to changes in the balance of economic forces and yields to the dictates of economic development. Thus, in the case of nineteenth-century Germany, the requirements of industrial production had become incompatible with feudal, politically fragmented Germany. Though political reaction was victorious in 1815 and again in 1848, it was unable to prevent the growth of large-scale industry in Germany and the growing participation of German commerce in the world market. In summary, Engles argued that "German unity had become an economic necessity."

In the view of both Smith and Engels the nation-state represented a progressive stage in human development because it enlarged the political realm of economic activity. In each successive economic epoch the advancing technology and scale of production necessitates an enlargement of political organization. Because the city-state and feudalism were below the optimum for the scale of production and the division of labor required by the Industrial Revolution, they prevented the efficient utilization of resources and were superseded by larger political units. Smith considered this to be a desirable objective; for Engels it was a historical necessity.

In contrast to the position of liberals and Marxists alike who stress the primacy of economic relations nationalists and the so-called realist school of political science have emphasized the primacy of politics. Whereas the liberal or Marxist emphasizes the production of wealth as the basic determinant of social and political organization, the realist stresses power, security, and national sentiment. . . .

Although himself a proponent of economic liberalism, the late Jacob Viner made one of the best analyses of the relationship of economic and political factors in determining the structure of international relations and concluded that political and security considerations are primary. In his classic study, *The Customs Union Issue,* Viner analyzed all known cases of economic and political unification from the perspective of whether the basic motivation was political or economic. Thus, whereas Engels interpreted the formation of the Zollverein as a response to the industrialization of Germany and the economic necessity of larger markets, Viner argued "that Prussia engineered the customs union primarily for political reasons, in order to gain hegemony or at least influence over the lesser German states. It was largely in order to make certain that the hegemony should be Prussian and not Austrian that Prussia continually opposed Austrian entry into the Union, either openly or by pressing for a customs union tariff lower than highly protectionist Austria could stomach." In pursuit of this strategic interest it was "Prussian might, rather than a common zeal for political unification arising out of economic partnership, [that] had played the major role."

Whereas liberalism and Marxism foresee economic factors leading to the decline of political boundaries and eventually to political unification, Viner argued that economic and political boundaries need not coincide and may actually be incompatible with one another. The tendency today, he pointed out, to take the identity of political and economic frontiers for granted is in fact a quite modern phenomenon and is even now not universal. With respect to tariffs, the concern of his study, the general rule until recently was that political unification was greater than the area of economic unification. Furthermore, any attempt to further economic unification might undermine political unification; this was the case with respect to the American Civil War and is the case today in Canada.

Viner concluded his argument that economic factors are of secondary importance to political unification with the following observation which is highly relevant for the concerns of this essay:

> The power of nationalist sentiment can override all other considerations; it can dominate the minds of a people, and dictate the policies of government, even when in every possible way and to every conceivable degree it is in sharp conflict with what seem to be and are in fact the basic economic interests of the people in question. To accept as obviously true the notion that the bonds of allegiance must necessarily be largely economic in character to be strong, or to accept unhesitatingly the notion that where economic entanglements are artificially or naturally strong the political affections will also necessarily become strong, is to reject whatever lessons past experience has for us in this field.

The contemporary argument that interstate relations will recede in face of contemporary technological developments and will be replaced by transnational relations between large multinational corporations was anticipated in the 1930s by Eugene Staley. In a fascinating book, *World Economy in Transition,* Staley posed the issue which is our main concern: "A conflict rages between technology and politics. Economics, so closely linked to both, has become the major battlefield. Stability and peace will reign in the world economy only when, somehow, the forces on the side of technology and the forces on the side of politics have once more been accommodated to each other."

While Staley believed, along with many present-day writers, that politics and technology must ultimately adjust to one another, he emphasized, in contrast to contemporary writers, that it was not inevitable that politics adjust to technology. Reflecting the intense economic nationalism of the 1930s, Staley pointed out that the adjustment may very well be the other way around. . . .

II

This . . . discussion of the relationship between economics and politics argues the point that, although the economic and technical substructure partially determines and interacts with the political superstructure, political values and security interests are crucial determinants of international economic relations. Politics determines the framework of economic activity and channels it in directions which tend to serve the political objectives of dominant political groups and organizations. Throughout history each successive hegemonic power has organized economic space in terms of its own interests and purposes.

Following in this vein, the thesis of this essay is that transnational actors and processes are dependent upon peculiar patterns of interstate relations. Whether one is talking about the merchant adventurers of the sixteenth century, nineteenth-century finance capitalists, or twentieth-century multinational corporations, transnational actors have been able to play an important role in world affairs because it has been in the interest of the predominant power(s) for them to do so. As political circumstances have changed due to the rise and decline of nation-states, transnational processes have also been altered or ceased altogether. Thus, . . . the world economy did not develop as result of competition between equal partners but through the emergence and influence of great national economies that successively became dominant.

From this perspective the multinational corporation exists as a transnational actor today because it is consistent with the political interest of the world's dominant power, the United States. This argument does not deny the analyses of economists who argue that the multinational corporation is a response to contemporary technological and economic developments. The argument is rather that these economic and technological factors have been able to exercise their profound effects because the United States—sometimes with the cooperation of other states and sometimes over their opposition—has created the necessary political framework. By implication, a diminution of the Pax Americana and the rise of powers hostile to the global activities of the multinational corporations would bring their reign over international economic relations to an end.

Perhaps the most effective way to defend the thesis that the pattern of international economic relations is dependent upon the structure of the international political system is to review the origins of the Pax Britannica, its demise with the First World War, and the eventual rise of a Pax Americana after the Second World War. What this history clearly reveals is that transnational economic processes are not unique to our own age and that the pattern of international economic activity reflects the global balance of economic and military power.

Each successive international system that the world has known is the consequence of the territorial, diplomatic, and military realignments that have followed history's great wars. The origins of the Pax Britannica lie in the complicated series of negotiations that followed the great upheavals of the Napoleonic wars. The essential features of the system which were put into place at that time provided the general framework of international economic relations until the collapse of the system under the impact of the First World War.

The first essential feature of the Pax Britannica was the territorial settlement and the achievement of a balance of power among the five Great Powers. This territorial realignment can be divided into two parts. In the first place, on the continent of Europe the territorial realignments checked the ambitions of Russia in the east and France in the west. Second, the overseas conquests of the continental powers were reduced at the same time that Great Britain acquired a number of important strategic overseas bases. As a result the four major powers on the Continent were kept in check by their own rivalries and by offshore Britain which played a balancing and mediating role.

British naval power, the second essential feature of the Pax Britannica, was able to exercise a powerful and pervasive influence over global politics due to a fortunate juncture of circumstances. Great Britain's geographical position directly off the coast of continental Europe and its possession of several strategic naval bases enabled it to control Europe's access to the outside world and to deny overseas colonies to continental governments. As a consequence, from 1825 when Great Britain warned France not to take advantage of the revolt of the Spanish colonies in America to the latter part of the century, the greater part of the non-European world was either independent or under British rule. Moreover, the maintenance of this global military hegemony was remarkably inexpensive; it thus permitted Great Britain to utilize its wealth and energies in the task of economic development.

Third, using primarily the instruments of free trade and foreign investment in this political-strategic framework, Great Britain was able, in effect, to restructure the international economy and to exercise great influence over the course of international affairs. As the world's first industrial nation, Great Britain fashioned an international division of labor which favored its own industrial strengths at the same time that it brought great benefits to the world at large. Exchanging manufactured goods for the food and raw materials of other nations, Great Britain was the industrial and financial center of a highly interdependent international economy.

One may reasonably argue, I believe, that in certain respects the regime of the Pax Britannica was the Golden Age of transnationalism. The activities of private financiers and capitalists enmeshed the nations in a web of interdependencies which certainly influenced the course of international relations. In contrast to our own era, in which the role of the multinational corporation in international economic relations is unprecedented, the private institutions of the City of London under the gold standard and the regime of free trade had a strategic and central place in world affairs unmatched by any transnational organization today. Prior to 1914 the focus of much of international relations was the City of London and the private individuals who managed the world's gold, traded in commodities, and floated foreign loans. Though this interdependence differs radically in kind from the internationalization of production and the immense trade in manufactured goods which characterize our own more industrialized world economy, this earlier great age of transnationalism should not be overlooked. . . .

The foundations underlying the Pax Britannica and the transnational processes it fostered began to erode in the latter part of the nineteenth century. On the Continent the industrialization and unification of Germany profoundly altered the European balance of power. France, too, industrialized and began to challenge Great Britain's global supremacy. Overseas development of equal or potentially greater magnitude were taking place. The rapid industrialization of Japan and the United States and their subsequent creation of powerful navies ended British control of the seas. No longer could Great Britain use its naval power to deny rivals access to the globe. With the decline of British supremacy the imperial struggle for the division of Africa and Asia began, leading eventually to the outbreak of the First World War.

The war completed the destruction of the pre-1914 system. As a consequence of the duration and intensity of the conflict one sector after another of economic life was nationalized and brought into the service of the state. The role of the state in economic affairs became pervasive, and economic nationalism largely replaced the laissez faire traditions upon which so much of pre-war transnationalism had rested. Not until the Second World War would political relations favor the reemergence of extensive transnational activity.

The failure to revive the international economy after the First World War was due to many causes: the policies of economic revenge against Germany; the ill-conceived attempt to reestablish the gold standard; the nationalistic "beggar-my-neighbor" policies pursued by most states, etc. In terms of our primary concern in this essay one factor in particular needs to be stressed, namely, the failure of the United States to assume leadership of the world economy, a role Great Britain could no longer perform. Whereas before the war the City of London provided order and coordinated international economic activities, now London was unable and New York was unwilling to restructure the international economy disrupted by the First World War. The result was a leadership vacuum which contributed in part to the onset of the Great Depression and eventually the Second World War.

For our purposes two developments during this interwar period hold significance. The first was the Ottawa Agreement of 1932 which created the sterling area of imperial preference and reversed Great Britain's traditional commitment to multinational free trade. The purpose of the agreement between Great Britain and the Commonwealth, an action whose intellectual roots went back to the nineteenth century, was to establish a regional trading bloc effectively isolated from the rest of the world economy. Germany in central Europe and Japan in Asia followed suit, organizing under their hegemonies the neighboring areas of strategic and economic importance. "This development of trading blocs led by great powers," one authority writes, "was the most significant economic development of the years immediately preceding the Second World War. As always the breakdown of international law and economic order gave opportunity to the ruthless rather than to the strong."[1] Such a system of law and order the international gold standard had provided. Under this system transnational actors could operate with little state interference. With its collapse nation-states struggled to create exclusive spheres of influence, and trade relations became instruments of economic warfare.

The second important development from the perspective of this essay was the passage of the Reciprocal Trade Agreements Act in June 1934. The purpose of this act was to enable the United States government to negotiate reductions in tariff barriers. Followed in 1936 by the Tripartite Monetary Agreement, the act not only reflected the transformation of the United States into a major industrial power but also represented the first step by the United States to assert its leadership of the world economy. Furthermore, it demonstrated the potential of bilateral negotiation as a method to achieve the expansion of multinational trade even though the immediate impact of the act was relatively minor. World trade continued to be dominated by preference systems, especially the sterling area, from which the United States was excluded. The importance of this prewar situation and the determination of the United States to overcome this discrimination cannot be too greatly emphasized. The reorganization of the world economy was to be the keynote of American postwar planning.

III

American plans for the postwar world were based on several important assumptions. In the first place, American leadership tended to see the origins of the Second World War as largely economic. The failure to revive the international economy after the First World War and the subsequent rise of rival trading blocs were regarded as the underlying causes of the conflict. Second, it was assumed that peace would be best promoted by the establishment of a system of multinational trade relations which guaranteed to all states equal access to the world's resources and markets. Third, the main obstacles to the achievement of such a universal system, Americans believed, were the nationalistic and discriminatory measures adopted in the 1930s by various European countries—trade preferences, exchange controls, quantitative restrictions, competitive currency depreciations, etc.

The importance of economic considerations in American postwar planning has led in recent years to a spate of writings by revisionist historians who interpret these efforts as part of a large imperial design. While this literature does serve to correct the simple-minded orthodox position that the cold war originated as a Communist plot to achieve world domination, it goes much too far and distorts the picture in another direction.

There is no question that the creation of a system of multilateral trade relations was in the interests of the United States. Preference systems ran directly counter to American basic interests as the world's dominant economic power and a major trading nation. It does not follow from this fact, however, that American efforts to achieve such a system were solely self-serving and unmotivated by the sincere belief that economic nationalism and competition were at the root of the Second World War. Nor does it follow that what is good for the United States is contrary to the general welfare of other nations.

The American emphasis on postwar economic relations represented a long tradition in American thought on international relations. The American liberal ideal since the founding of the Republic has been the substitution of commercial for political relations between states. In the best free trade tradition, trade relations between nations are considered to be a force for peace. Furthermore, as a nation which felt it had been discriminated against by the more powerful European states, the United States wanted a world in which it would have equal access to markets. Universal equality of opportunity, not imperial domination, was the motif of American postwar foreign economic planning.

This naive American faith in the beneficial effects of economic intercourse was reflected in the almost complete absence of attention to strategic matters in American postwar plans. In contrast to the prodigious energies devoted to the restructuring of the international economy little effort was given to the strategic and territorial balance of the postwar world. This neglect is explainable in large part, however, by the prevailing American assumption that a universal system based on an integrated world economy and on the United Nations would replace the traditional emphasis on spheres of influence and the balance of power.

If one accepts the revisionist argument that imperial ambition underlay American postwar plans, then the cold war should have been between the United States and Western Europe, particularly the United Kingdom, rather than between the Union of Soviet Socialist Republics and the United States. The bete noir of American planners was European discrimination and especially the imperial preference which encompassed a high percentage of world trade and exercised considerable discrimination against American goods. American plans for the postwar era were directed against the British in particular. Beginning with the framing of the Atlantic Charter in 1941 and continuing through the negotiation of the Lend-Lease Act (1941), the Bretton Woods Agreement (1944), and the British Loan (1945), the thrust of American policy was directed against Commonwealth discrimination.

In light of the intensity of these American efforts to force the United Kingdom and other European countries to accept a multilateral system it is important to appreciate that they were abandoned in response to growth of Soviet–American hostility. As American leadership came to accept the Soviet diplomatic-military challenge as the major postwar problem, the United States attitude toward international economic relations underwent a drastic reversal. In contrast to earlier emphases on multilateralism and nondiscrimination the United States accepted discrimination in the interest of rebuilding the shattered West European economy.

The retort of revisionists to this argument is that the American–Soviet struggle originated in the American desire to incorporate Eastern Europe, particularly Poland, into the American scheme for a global empire. This effort, it is claimed, clashed with the legitimate security concerns of the Soviet Union, and the cold war evolved as the Soviet defensive response to the American effort to expand economically into the Soviet sphere of influence. If the United States had not been driven by the greed of its corporations, American and Soviet interests could easily have been accommodated.

There are sufficient grounds for this interpretation to give it some plausibility. Certainly, American efforts to incorporate Eastern Europe and even the Soviet Union into the world capitalistic economy raised Soviet suspicions. Although the American view was that the withdrawal of the Soviet Union from the world economy following the Bolshevik Revolution had been a contributing factor to the outbreak of the Second World War and that a peaceful world required Soviet reintegration, the Russians could easily interpret these efforts as an attempt to undermine communism. No doubt in part they were. But it is a long jump from these American efforts to trade in an area of little historical interest to the United States to a conflict so intense and durable that it has on several occasions taken the world to the brink of thermonuclear holocaust.

A more realistic interpretation, I believe, is that the origins of the cold war lie in the unanticipated consequences of the Second World War. The collapse of German power in Europe and of Japanese power in Asia created a power vacuum which both the United States and the Soviet Union sought to fill to their own advantage. One need not even posit aggressive designs on either side to defend this interpretation, although my own position is that the Soviet Union desired (and still desires) to extend its sphere of influence far beyond the glacis of Eastern Europe. To support this political interpretation of the cold war it is sufficient to argue that the power vacuums in Central Europe and the northwestern Pacific created a security dilemma for both powers. In terms of its own security neither power could afford to permit the other to fill this vacuum, and the efforts of each to prevent this only increased the insecurity of the other, causing it to redouble its own efforts. Each in response to the other organized its own bloc, freezing the lines of division established by the victorious armies and wartime conferences.

One cannot understand, however, the pattern of the cold war and its significance for international economic relations unless one appreciates the asymmetric situations of the United States and the Soviet Union. Whereas the Soviet Union is a massive land power directly abutting Western Europe and the northwestern Pacific (primarily Korea and Japan), the United States in principally a naval and air power separated from the zones of contention by two vast oceans. As a consequence, while the Soviet Union has been able with relative ease to bring its influence to bear on its periphery at relatively much less cost in terms of its balance of payments, the United States has had to organize a global system of bases and alliances involving an immense drain on its balance of payments. Moreover, while the Soviet system has been held together largely through the exercise of Soviet military power, economic relations have been an important cement holding the American bloc together.

These economic and strategic differences between the two blocs have been crucial determinants of the postwar international economy and the patterns of transnational relations which have emerged. For this reason some attention must be given to the interplay of economic and political factors in the evolution of relations between the three major components of the contemporary international economy: the United States, Western Europe, and Japan.

Contrary to the hopes of the postwar economic planners who met at Bretton Woods in 1944, the achievement of a system of multilateral trade was soon realized to be an impossibility. The United Kingdom's experience with currency convertibility, which had been forced upon it by the United States, had proven to be a disaster. The United Kingdom and the rest of Europe were simply too weak and short of dollars to engage in a free market. A further weakening of their economies threatened to drive them into the arms of the Soviet Union. In the interest of preventing this the United States in cooperation with Western Europe had to rebuild the world economy in a way not envisaged by the postwar planners.

The reconstruction of the West European economy involved the solution of three problems. In the first place, Europe was desperately short of the dollars required to meet immediate needs and to replenish its capital stock. Second, the prewar European economies had been oriented toward colonial markets. Now the colonies were in revolt, and the United States strongly opposed the revival of a world economy based on a colonial preference system. Third, the practices of economic nationalism and closed preference systems between European states and their overseas colonies had completely fragmented the European economy.

The problem of rehabilitating the economy of the Federal Republic of Germany (West Germany) was particularly difficult. The major trading nation on the Continent, its division into Soviet and Western zones and the Soviet occupation of Eastern Europe had cut industrial West Germany off from its natural trading partners in the agricultural German Democratic Republic (East Germany) and the East. The task therefore was to integrate the industrial Western zones into a larger West European economy comprising agricultural France and Italy. The failure to reintegrate industrial Germany into the larger world economy was regarded to have been one of the tragic errors after World War I. A repetition of this error would force West Germany into the Soviet Camp.

The American response to this challenge is well known. Through the Marshall Plan, the Organization for European Economic Cooperation (OEEC), and the European Coal and Steel Community (ECSC) the European economy was revived and radically transformed. For our purposes one point is significant. In the interest of security the United States tolerated, and in fact promoted, the creation of a preference area in Western Europe which discriminated against American goods. At first the mechanism of discrimination was the nonconvertibility of European currencies; then, after the establishment of the European Economic Community (EEC) in 1958, discrimination took the form of one common external tariff.

The economic impact of economic regionalism in Western Europe was not, however, completely detrimental to United States–European trade. One can in fact argue that regionalism gave Europe the courage and security to depart from traditions of economic nationalism and colonialism. The establishment of a large trading area in Europe turned out to be more trade-creating than trade-diverting. As a consequence American and European economic ties increased and the United States continued to enjoy a favorable balance of trade with its European partners.

With respect to Japan the United States faced a situation similar to that presented by West Germany. Although Japan was not severely damaged by the war, it was a densely populated major trading nation exceptionally dependent upon foreign sources of raw materials, technology, and agricultural products. With the victory of the communists on the Chinese mainland Japan's major prewar trading partner came under the control of the Soviet bloc. Furthermore, Japan suffered from discrimination by other industrialized states both in their home markets and in their overseas colonial empires. The exclusion of the Japanese from South and Southeast Asia practiced by the Dutch, French, and British had been a major cause of Japan's military aggression, and the continued existence of these preference systems threatened its economic well-being. Separated from the Soviet Union by a small body of water and economically isolated, Japan's situation was a highly precarious one.

As in the case of West Germany the task of American foreign Policy was to integrate Japan into the larger world economy and lessen the attraction of markets controlled by the Communist bloc. While this history of American efforts to restructure Japan's role in the world economy is less well known than is the history of its European counterpart, the basic aspects deserve to be emphasized. In the first place, the United States brought pressures to bear against Dutch, French, and British colonialism in South and Southeast Asia and encouraged the integration of these areas into a larger framework of multilateral trade. Second, over the strong opposition of Western Europe the United States sponsored Japanese membership in the International Monetary Fund (IMF), the General Agreement on Tariffs and Trade (GATT), and other international organizations. Third, and most significant, the United States in the negotiations leading to the Treaty of Peace with Japan granted Japan privileged access to the American home market.

At the same time that these developments in the economic realm were taking place, through the instrumentalities of the North Atlantic Treaty Organization (NATO) and the Treaty of Peace with Japan, Western Europe and Japan were brought under the protection of the American nuclear umbrella. In Europe, Japan, and around the periphery of the Soviet Union and the People's Republic of China (Communist China) the United States erected a base system by which to counter the Soviet advantage of geographical proximity. Thus, with their security guaranteed by this Pax Americana, Japan, Western Europe, and, to a lesser extent, the United States have been able to devote the better part of their energies to the achievement of high rates of economic growth within the framework of a highly interdependent transnational economy.

Just as the Pax Britannica provided the security and political framework for the expansion of transnational economic activity in the nineteenth century, so this Pax Americana has fulfilled a similar function in the mid-twentieth century. Under American leadership the various rounds of GATT negotiations have enabled trade to expand at an unprecedented rate, far faster than the growth of gross national product in the United States and Western Europe. The United States dollar has become the basis of the international monetary system, and, with the rise of the Eurodollar market, governments have lost almost all control over a large segment of the transnational economy. Finally, the multinational corporation has found the global political environment a highly congenial one and has been able to integrate production across national boundaries.

The corollary of this argument is, of course, that just as a particular array of political interests and relations permitted this system of transnational economic relations to come into being, so changes in these political factors can profoundly alter the system and even bring it to an end. If, as numerous writers argue, there is a growing contradiction between the nation-state and transnational activities, the resolution may very well be in favor of the nation-state or, more likely, of regional arrangements centered on the dominant industrial powers: Japan, the United States, and Western Europe.

I V

This argument that contemporary transnational processes rest on a peculiar set of political relationships can be substantiated, I believe, if one analyzes the two most crucial relationships which underlie the contemporary international economy. The first is the relationship between the United States and West Germany, the second is that between the United States and Japan.

While the American–West German special relationship is based on a number of factors including that of mutual economic advantage, from the perspective of transnational activities one factor is of crucial importance. In simplest terms this is the exchange of American protection of West Germany against the Soviet Union for guaranteed access to EEC markets for American products and direct investment. . . . With respect to direct investment the subsidiaries of American corporations have been able to establish a very powerful position in Western Europe since the beginning of the EEC in 1958.

Without . . . West German willingness to hold dollars, the American balance-of-payments situation might, the West Germans fear, force the United States to reduce its troop strength in West Germany. As such a move could lessen the credibility of the American nuclear deterrent, the West Germans are very reluctant to make any moves which would weaken the American presence in Western Europe. Consequently, while the significance of American direct investment in Europe for the American balance of payments is unclear, the West Germans are unwilling to take any action regarding this investment which might alienate American opinion and lessen the American commitment to Western Europe.

Turning to the other pillar of the contemporary transnational economy the American–Japanese special relationship, mutual economic interest is an important bond, but the primary factor in this relationship has been the security issue. In contrast to the American–West German situation, however, this relationship involves American protection and a special position for the Japanese in the American market in exchange for United States bases in Japan and Okinawa. The asymmetry of this relationship compared with that between the United States and West Germany reflects the differences in the economic and military situations.

As mentioned earlier the basic problem for American foreign policy with respect to Japan was how to reintegrate this highly industrialized and heavily populated country into the world economy. Given communist control of mainland Asia and the opposition of European countries to opening their markets to the Japanese this meant throwing open the American economy to Japanese exports. As a consequence of this favored treatment the Japanese have enjoyed an exceptionally favorable balance of trade with the United States. . . .

In contrast to the situation prevailing in Europe the purpose of American military base structure in Japan is not merely to deter local aggression against the Japanese; rather, it is essential for the maintenace of American power and influence throughout the western Pacific and Southeast Asia. Without access to Japanese bases the United States could not have fought two wars in Asia . . . and could not continue its present role in the area. . . .

In the case of both the American–European and the American–Japanese relationships new forces are now at work which threaten to undermine the foundations of contemporary transnational relations. In the case of United States–European relations the most dramatic change is the decreased fear of the Soviet Union by both partners. As a consequence both Americans and Europeans are less tolerant of the price they have to pay for their special relationship. The Europeans feel less dependent upon the United States for their security and are more concerned with the detrimental aspects of close economic, military, and diplomatic ties with the United States. The United States, for its part, is increasingly sensitive to European discrimination against American exports and feels threatened by EEC moves toward the creation of a preference system encompassing much of Western Europe, the Middle East, and Africa. . . .

With respect to the relationship of Japan to the United States, strategic and economic changes are undermining the foundations of transnationalism. At the same time that Communist China is receding as a security threat to the United States and Japan, economic strains are beginning to aggravate relations between the two countries. In the eyes of the United States Japan's economy is no longer weak and vulnerable, necessitating special consideration by the United States. As a consequence the demands of American interests for import curb against Japanese goods and for the liberalization of Japanese policies on foreign direct investment are beginning to take precedence over foreign policy and strategic considerations. Nor does the United States continue to accept the fact that the defense burden should rest so heavily on it alone. Underlying the Nixon Doctrine of American retrenchment in Asia is the appreciation that a greater Japanese military effort would not only reduce American defense costs but would also cause the Japanese to divert resources from their export economy and relieve Japanese pressures in the American market.

The Japanese for their part resent the fact that they are almost totally dependent upon the United States for their security and economic well-being. While they of course want to maintain a strong position in the American market and feel particularly threatened by protectionist sentiment in the United States, they are growing increasingly concerned about the price they must pay for their close association with the United States. Moreover, they feel especially vulnerable to American economic pressures such as those that have been exerted to induce Japan to permit direct investment by American corporations. But the dominant new factor is the Japanese desire to play a more independent role in the world and to enjoy the prestige that is commensurate with their powerful and expanding economy.

In the cases of both American–European and American–Japanese relations new strains have appeared which threaten to undermine the political framework of transnational economic activity. Diplomatic and military bonds tying Europe and Japan to the United States have weakened and at the same time that economic conflicts have intensified and have become less tolerable to all three major parties. As a result the favorable political factors that have facilitated the rapid expansion of transnational processes over the past several decades are receding. In their stead new political forces have come into play that are tending to isolate the United States and to favor a more regional organization of the international economy.

On the other hand, one must readily acknowledge that the multinational corporation and transnational processes have achieved tremendous momentum. It is not without good reason that numerous authorities have predicted the demise of the nation-state and the complete reordering of international life by 200 or 300 "megafirms." Perhaps, as these authorities argue, the multinational corporation as an institution has sufficiently taken root in the vested interests of all major parties that it can survive the vicissitudes of political change. History, however, does not provide much comfort for this train of thought. As Staley and Viner have suggested, the contradiction between the economic and political organization of society is not always resolved in favor of economic rationality. Moreover, whatever the outcome—the preservation of multilateral transnational processes, a reversion to economic nationalism, or the division of the globe by economic regionalism—the determining consideration will be the diplomatic and strategic interests of the dominant powers.

V

Prior to concluding this essay one crucial question remains to be treated: What, after all, has been the impact of transnational economic activities, especially the multinational corporation, on international politics? In answer to this question both Marxists and what one might call the transnational ideologists see these transnational processes and actors as having had a profound impact on international relations. Some go much further. By breaking the monopoly of the nation-state over international economic relations the multi-national corporation is claimed to have altered the very nature of international relations.

Under certain circumstances and in relation to particular states there can be little doubt that the multinational corporation has, and can exercise, considerable influence over domestic and international relations. One could mention in this connection the international petroleum companies, for example. But in general there is little evidence to substantiate the argument that the multinational corporation as an independent actor has had a significant impact on international politics. As Staley has convincingly shown in his study of foreign investment prior to World War II, where business corporations have exercised an influence over political developments they have tended to do so as intruments of their home governments rather than as independent actors.

Contemporary studies on the multinational corporation indicate that Staley's conclusion continues to hold true. While the evidence is indisputable that the multinational corporation is profoundly important in the realm of international economic relations, its political significance is largely confined to its impact on domestic politics where it is an irritant to nationalistic sentiments. In part the resentment has been due to the unwarranted interference by foreign-owned corporations in domestic affairs; this has especially been the case in less developed countries. More frequently, nationalistic feelings have been aroused by the predominant positions multinational corporations may hold in the overall economy or in particularly sensitive sectors.

Despite all the polemics against multinational corporations there is little evidence to support the view that they have been very successful in replacing the nation-state as the primary actor in international politics. Where these business enterprises have influenced international political relations, they have done so, like any interest group, by influencing the policies of their home governments. Where they have tried to influence the foreign and economic policies of host governments, they have most frequently been acting in response to the laws of their home countries and as agents of their home governments. . . .

Contrary to the argument that the multinational corporation will somehow supplant the nation-state, I think it is closer to the truth to argue that the role of the nation-state in economic as well as in political life is increasing and that the multinational corporation is actually a stimulant to the further extension of state power in the economic realm. One should not forget that the multinational corporation is largely an American phenomenon and that in response to this American challenge other governments are increasingly intervening in their domestic economies in order to counter balance the power of American corporations and to create domestic rivals of equal size and competence.

The paradox of the contemporary situation is that the increasing interdependence among national economies, for which the multinational corporation is partially responsible, is accompanied by increased governmental interference in economic affairs. What this neo-mercantilism constitutes, of course, is one response to the basic contradiction between the economic and political organization of contemporary international society. But in contrast to the opinion of a George Ball who sees this conflict resolved in favor of transnational processes, the internationalization of production, and actors like the multinational corporation, nationalists in Canada, Western Europe, and the less developed world favor upholding more powerful states to counterbalance large multinational corporations.

Similarly, the impetus today behind the EEC, Japan's effort to build an economic base less dependent on the United States, and other moves toward regionalism reflect in part a desire to lessen the weight of American economic power; in effect, these regional undertakings are essentially economic alliances between sovereign governments. Although they are altering the political framework within which economic forces will increasingly have to operate, the basic unit is and will remain the nation-state. For better or for worse it continues to be the most powerful object of man's loyalty and affection.

Notes

1. J. B. Condliffe, *The Commerce of Nations* (New York: W. W. Norton & Co. 1950), p. 502.

Domestic and International Forces and Strategies of Foreign Economic Policy

PETER J. KATZENSTEIN

The management of interdependence is a key problem which all advanced industrial states have confronted in the postwar international political economy. Differences in their domestic structures and the international context in which they are situated have dictated the adoption of different strategies of foreign economic policy. The rationale of all strategies is to establish a basic compatibility between domestic and international policy objectives. But since the domestic structures in the advanced industrial states differ in important ways, so do the strategies of foreign economic policy which these states pursue. . . .

In the contemporary international political economy increasing interdependence goes hand in hand with assertions of national independence. As was true of military policy in the previous era of "national security," in the present era of "international interdependence," strategies of foreign economic policy depend on the interplay of domestic and international forces. The starting premise of this volume thus puts little value on a clear-cut distinction between domestic and international politics. A selective focus on either the primacy of foreign policy and the "internationalization" of international effects or on the primacy of domestic politics and the "externalization" of domestic conditions is mistaken. Such a selective emphasis overlooks the fact that the main purpose of all strategies of foreign economic policy is to make domestic politics compatible with the international political economy.

But the literature on foreign economic policy has, in recent years, unduly

Reprinted from *International Organization*, Vol. XXXI, No. 4, Autumn 1977, Peter J. Katzenstein, "Introduction: Domestic and International Forces and Strategies of Foreign Economic Policy," by permission of the MIT Press, Cambridge, MA.

discounted the influence of domestic forces. Explanations which focus on the persistence of international interdependence and the pervasiveness of transnational relations have failed to account for a paradox in our understanding of the international political economy. These international explanations do not adequately explain why an international challenge, such as the oil crisis, elicits different national responses. Despite the enormous growth of different forms of international interdependencies and transnational relations, the nation-state has reaffirmed its power to shape strategies of foreign economic policy. . . .

. . . . True to Hegel's maxim, events rather than intellectuals have forced this reorientation in theoretical perspectives . . . The Vietnam War, it turned out, was not the product of strategic interaction between the superpowers but of Washington's institutionalization of a world order infused by an encompassing structure of "interest and ideology." In the 1960s, the foreign policies of the European states also illuminated the importance of domestic structures. General de Gaulle shifted the course of European politics by obstinately refusing to succumb to the force of European integration. And Japan's stunning rise to the position of one of the world's foremost commercial powers was evidently made possible by particular features in its domestic politics. In the 1970s, it is not possible to understand the international energy system without appreciating the domestic barriers which impede an American energy policy. Similarly, the persistent and enormous surplus in the West German trade balance (and the burden which it puts on other deficit countries) can be understood only when one knows of the strong forces in West German politics which are opposed to inflationary policies. The decline in America's influence in world politics is countered by the growing political importance of other nation-states. Today's international political economy remains unintelligible without a systematic analysis of domestic structures.

I. Why Study The "North"?

When we think about problems of the international political economy our imagination is captured by the current dialogue between North and South. Negotiations in Paris, Nairobi, and New York have wavered between confrontation and compromise and appear to be making only very slow progress. While the North will consider only marginal adjustments, the South insists on far-reaching modifications of present international economic arrangements. This difference is encapsulated in the political rhetoric of the two parties to the negotiations. The South speaks of *The* New International Economic Order; the North refers to it, typically, as *A* New International Economic Order. In asking for a redistribution of resources the South raises matters of substance. In talking about instrumentalities and procedures, the North has responded, until recently, by raising matters of style. The idea of a new "international class conflict" between North and South is intuitively plausible to many and may, over the long run, have important political consequences. But for the short- and medium-term it is important to separate fact from fiction. The notion that a new international class conflict could determine developments in the international political economy during the years ahead is based on two questionable political assumptions.

First, the image of bloc conflict between North and South presumes a degree of international order which is lacking today. Both politically and economically the two camps are internally divided. There are vast differences, for example, in the economic circumstances of oil-producing states such as Saudi Arabia, Iran, Indonesia, Venezuela or, soon, Britain and Norway. They have different balance-of-payment objectives, growth potentials, and export dependencies. Differences of similar magnitude can be observed in the domestic social and political structures which condition the foreign economic policies of these states. Among the producer countries cartelization, it appears, is difficult; oil will be, for some time to come, "the exception." The political and economic heterogeneity of raw-material exporters is a formidable barrier inhibiting the organization of cartels.

Among the advanced industrial states these differences in economic and political circumstances are equally striking. The economic crisis faced by Britain and Italy, for example, is much more serious than those of other states in the North. On the surface, differences in domestic social and political structures may be less glaring than among the countries of the South. But . . . important differences affect foreign economic policy. These show little sign of disappearing. In sum, the image of bloc politics in the international political economy, of the confrontation between producer and consumer cartels, is deceptively simple and misleading.

There is a second reason why the image of bloc politics between North and South is misplaced. In the past, major shifts in the distribution of power have occurred in domestic politics when the interlocking tripod of power, wealth, and status began to split not from the bottom but from the top. The disorder in the current international political economy did not originate with the oil embargo in 1973 but dates back to the mid-1960s and the growing strains in the international monetary system. These strains resulted from the redistribution of economic power between the reconstructed economies of the European continent and Japan, on the one hand, and of the United States on the other. Increasing trade conflicts, especially in the area of agriculture, and growing antagonisms toward overseas investment of American multinational corporations were noticeable in the late 1960s. More recently, the imposition of the oil embargo and the diverging reactions of the rich countries illustrated the deep fissures which prevented the advanced industrial states from meeting the crisis in a coordinated fashion.

Despite the lack of coordination, the advanced industrial states still dominate the international political economy. Roughly two-thirds of global trade and direct investment is exchanged among a handful of advanced industrial states. In an era of rising concern over adequate international grain reserves, an overwhelming proportion of agricultural exports comes from the North American breadbasket. Only a small number of advanced industrial states has the technological capability for exporting nuclear reactors—a possible partial remedy for the global energy shortage and a likely stimulant to the proliferation of nuclear weapons. . . .

II. International And Domestic Forces And Foreign Economic Policy

Strategies of foreign economic policy of the advanced industrial states grow out of the interaction of international and domestic forces. That interaction is evident in the cycle of hegemonic ascendance and decline which we can trace in the international political economy over the last 150 years. An open international political economy favoring the process of international exchange existed during Britain's hegemonic ascendance (1840–1880) and during the prominence of the United States after World War II. These periods were followed by movements toward international closure in periods of hegemonic decline marked by numerous challenges to the established institutions governing foreign economic policy. This was true of the late nineteenth century and the interwar years; some elements of this development can also be detected since the late 1960s. Periods of imperial ascendance are distinguished by the politics of plenty. A coalition within a rising hegemonic state is able to maintain an open international political economy. State power is largely invisible, for the nature of the problem in international economic affairs centers on distribution and regulation which occur within established structures. The international political economy thus is orderly; this in turn facilitates the task of political management. Periods of hegemonic decline, on the other hand, are marked by the "politics of scarcity." A coalition within a declining hegemonic power is, eventually, neither willing nor able to resist the forces pushing toward closure. State power becomes visible, for the nature of the political problem in the international economy centers on redistribution and constitutional debates which question established institutions. The lack of order in the international political economy in turn impedes the exercise of effective political leadership by the declining hegemonic state.

From the 1820s on, Britain's industrial prowess and its decision to shift gradually toward a low-tariff policy made it the leading state in the international political economy of the nineteenth century. That role reflected a realignment in Britain's domestic balance of power which gradually tilted in favor of its commercial, industrial, and financial elites. The defeat which Britain's aristocratic landowners suffered in the repeal of the Corn Laws in 1846 was a critical divide in British history. Until Britain's entry into the Common Market in the early 1970s the principle of no tariffs on food imports would not be questioned again. But the repeal of the Corn Laws also had important consequences for the international political economy, for it opened Britain's market to the Continent's grain exports, thus facilitating international trade. The interests of Britain's rapidly expanding industrial, commercial, and financial sectors, cast in the ideology of Manchester Liberalism, received their international sign of legitimation with the signing of the Cobden-Chevalier Tariff Treaty of 1860, which set the stage for numerous other bilateral tariff reductions throughout Europe. A realignment of political forces within Britain was, then, projected onto the international political economy, where Britain's preeminence assured order.

In comparison both to the United States (which decided not to assume its role as political leader until 1945) and to Germany (which became Britain's main political rival in the late nineteenth century) Britain's hegemonic power weakened from the 1880s onward. This was illustrated by its unwillingness and inability to counteract successfully the growing wave of protectionism which spread from the late 1870s in response to the Depression of 1873–1896. Like economic liberalism, economic nationalism was the result of shifting coalitions in domestic politics. In Germany, for example, the (in)famous coalition between iron and rye was based on a high tariff policy and amounted to what has been called the Second Founding of the Empire. This realignment of political forces was as important for Germany as the Corn Laws had been for Britain, and it was critical in pushing the international political economy off its liberal course. Germany's challenge to Britain eventually led to an intensive politicization of the international economy, as economic and military interests became insolubly linked both at home (as in the naval arms race) and abroad (as on the Baghdad Railway).

The absence of a hegemonic power and a protectionist international political economy was still more prominent during the Great Depression of the 1930s. The flow of goods, capital, and labor was regulated not by market forces but by states engaging in competitive protectionism and devaluation policies, making the interwar international economy largely political. The breakdown of a liberal international economic order reflected the profound social upheaval which World War I had wrought on all of Europe. The weakening of the old middle class and the intensification of the conflict between organized capital and organized labor pushed more European states toward the Right. Fascist and authoritarian ideologies legitimized the pervasive role of the state in economic affairs, both domestic and international.

With the United States acting the part of international leader, the cycle of hegemonic ascendance and decline has repeated itself since 1945. The Bretton Woods system, reinforced by free-convertibility achieved in 1959 and successive tariff reductions culminating in the Kennedy Round (completed in 1967), facilitated a degree of openness in the international political economy and a growth in international transactions unprecedented in the twentieth century. The multinational corporation, as the institutionalization of this liberal marketplace ideology, became the most dramatic symbol of an international political economy functioning smoothly under U.S. auspices. This international economic order was based on a convergence of domestic and foreign policy interests on both sides of the Atlantic. An open, depoliticized international market was perceived to be an essential precondition for the successful reestablishment of a "bourgeois Europe." Supported by the flow of U.S. government aid and private capital, the rapid economic recovery of the war-devastated states of Western Europe and Japan has led to a gradual diminution of the United States' hegemonic position since the mid-1960s. The oil crisis, however, halted the process of hegemonic decline vis-à-vis Western Europe. Through the adroit political maneuvering of Secretary of State Kissinger, the lesser dependence of the United States on raw material imports has temporarily arrested a further decline of the United States in international economic affairs. It remains to be seen whether the domestic stalemate on American energy policy will accelerate that decline in the near future.

International and domestic factors have been closely intertwined in the historical evolution of the international political economy since the middle of the nineteenth century. Shifts in domestic structures have led to basic changes in British, German, and American strategies of foreign economic policy. The international context in which these countries found themselves in turn influenced their domestic structures and thus, indirectly, the strategies they adopted in the international political economy. But the relative weight of domestic structures in the shaping of foreign economic policy increased in periods of hegemonic decline. As long as the distribution of power in the international political economy was not in question, strategies of foreign economic policy were conditioned primarily by the structure of the international political economy. But when that structure could no longer be taken for granted, as is true today, the relative importance of domestic forces in shaping foreign economic policy increased. Over the last decade the gradual shift from security issues to economic concerns has further increased the relative weight of domestic structures on foreign economic policy. Everywhere the number of domestic interests tangibly affected by the international political economy is far greater than it had been during the previous era of national security policy.

III. International And Bureaucratic Politics Approaches To The International Political Economy

Our focus on the effect of domestic structures on foreign economic policy complements two other approaches enjoying wide currency in contemporary writings on international political economy. One approach focuses on different political, economic, and technological features of the international system which affect strategies of foreign economic policy. The second approach examines in detail domestic, bureaucratic factors and the policy-making process. These two approaches have different strengths. A focus on international developments is particularly useful for analyzing the range of choice of foreign economic policy strategies. An analysis of bureaucratic factors highlights the contingency of strategies. . . .

International Approaches

Recent analyses have begun to distinguish systematically between the different international forces conditioning foreign economic policy. Derived from Realist, Marxist, and Liberal traditions, these analyses offer different explanations of the limits which the growing international division of labor has imposed on strategies of foreign economic policy. Some of the more recent analyses of transnational relations have attempted to construct eclectic explanations drawing on several of these intellectual traditions.

Despite the increase in the international division of labor and the growth in international commerce and finance, political realists argue that America's hegemonic presence in the international political economy has affected the strategies of foreign economic policy of all advanced industrial states since 1945. That hegemony rested on America's relative military and economic invulnerability, which has been evident even in recent years, for example, in the ending of the Bretton Woods system in 1971 and in the oil crisis of 1973. Marxist interpretations of the international political economy have viewed the internationalization of industrial and financial capital as both a reflection of and a constraint on political strategies. Both the internationalization of industrial capital through multinational corporations with "global reach" and the internationalization of financial capital through a variety of banking institutions are rooted in the structure of capitalist economies, limiting the range of political choices. Liberal interpretations, finally, have focused on the effects of technological change on increases in capital flows which reveal the sensitivity of national economies to international developments and also limit strategies. . . .

Although these three interpretations highlight different features of the international political economy, their particular strength lies in focusing attention primarily on the different kinds of limitations which the world economy imposes. These interpretations are, however, not so helpful in explaining the different strategies which advanced industrial states actually pursue. This is the central objective of this volume. Its primary focus on domestic structures facilitates an explanation of the different strategies; emphasis on international developments, on the other hand, often leaves unspecified the policy consequences of particular international developments for different advanced industrial states. In the future, though, it will be necessary to develop the connections between limitations on strategies and actual policy choices more systematically than has been possible in this volume.

Bureaucratic Politics

A growing number of books and articles have also focused on the machinery of government which conditions strategies of foreign economic policy. This literature is derived from Max Weber's ideal-typical concept of "rational" modern bureaucracy on the one hand, and the description of "irrational" bureaucracy in American organization theory on the other. The "rational" and the "irrational" types define the end points of a continuum along which different models of foreign policy and foreign economic policy have been constructed.

The American state has been the prime testing ground for the development of this approach to the study of foreign policy making. The complexity of the foreign policy machinery of the modern American state can be illustrated even in times of extreme political crisis. The rationality of bureaucracy imputed by the Weberian ideal type explains certain actions of the United States government, but it leaves many others unaccounted for. The size of this bureaucracy and the scale of its operations have become so large that they lead to a lack of control by central foreign policy decision makers. Informal networks and alliances, reinforced by idiosyncratic factors tied to the character and position of particular individuals, work in the same direction. With the shift of power from Congress to the Executive, interest-group politics has simply moved too. The rule of law in democratic systems gives the bureaucracy a larger autonomy from the executive than it might enjoy in other forms of government, thus further encouraging a political process within the bureaucracy.

Models of bureaucratic politics are useful in detailing the intra-bureaucratic limitations which the contingency of numerous complex factors imposes on strategies of foreign economic policy. These models are, however, less helpful in accounting for the different strategies pursued by the advanced industrial states. It is simply not clear to what extent the bureaucratic politics approach universalizes a particular American syndrome (the internal fragmentation of the American state) and how much it particularizes a universal phenomenon (the push and pull which accompany policy making in all advanced industrial states). Models of bureaucratic politics abstract so little from the different strategies of foreign economic policy which need to be explained that they cannot clearly specify either the content of political strategies or the influence of particular intra-bureaucratic factors in the advanced industrial states. In the future, though, it will perhaps be possible to develop a way of thinking about bureaucratic limitation which can be linked systematically to the analysis of domestic structures lying at the heart of this volume.

IV. State, Society, And Foreign Economic Policy

The essays in this volume share a theoretical orientation which we should make explicit at the outset. This common view includes both the problems in foreign economic policy worth explaining (the dependent variables) and the political and economic forces operating at home and abroad which make these problems intelligible (the independent variables).

In characterizing the problem one could focus on the specific decisions and events which constitute particular episodes in policy making. Or one could look at general features which characterize types of policies. In taking the first route, the case-study approach generates useful and important descriptive detail on specific phases of the policy process; but it often leaves the broader meaning of policy untouched. Discussion of the latter is the particular strength of the second approach. But comparative political research faces the problem of linking particular policies or phases of policy making to abstract definitions of different types of policies.

Between these two approaches [is] a middle course. Two components of strategies of foreign economic policy are singled out for analysis. The first is the definition of policy objectives or the ends of policy; the second is the instrument or means of policy. This distinction between objectives and instruments is an arbitrary analytical device; in many instances the defense of a means can become an end in itself, and ends can become means in the attainment of other objectives. In the case of British monetary policy, for example, it is virtually impossible to decide whether the defense of sterling was a means to defend the reserve status of the British currency, and with it Britain's international position, or whether it had become an end in itself. . . .

The key problem for a comparative analysis is to understand the differences between the advanced industrial states. In *Social Origins of Dictatorship and Democracy,* Barrington Moore postulated a democratic Liberal, a Fascist, and a Communist road to the present. However, today we find it hard to understand fully the varieties of domestic structures which Moore, emphasizing their similarities instead of their differences, referred to as "democractic." These varieties are analyzed here using the traditional distinction between "state" and "society."

In the history of political thought there is a strand of theory associated with the rise of the absolutist state. The writings of Machiavelli or Bodin focus mainly on statecraft and discount civil society as a passive object for the instruments of rulers. During the Enlightenment, writers began to focus attention on the social conditions which facilitated or impeded the exercise of statecraft under different constitutional regimes. This in fact was one of Montesquieu's purposes in the *Spirit of the Laws.* In the nineteenth century this evolved into a sociological perspective on the state which became a core element of both Liberal and Marxist interpretations of politics. In addition to these two contrasting perspectives a third emerged which focused on the state and on society as partly interdependent and partly autonomous spheres of action. The intellectual giants of the nineteenth century—Tocqueville, Marx, Durkheim, and Weber—grappled with the problem of how state and society counterbalanced each other in different historical periods and in different places.

Although this dualism between "state" and "society" was postulated before and after him—for example in the writings of Rousseau and Weber—Hegel's distinction between the public realm of freedom and the private realm of necessity was the cornerstone of his theory of the state and his philosophy of history. Hegel's civil society strongly resembled the economic marketplace of Manchester Liberalism, for it was constituted of the complex interactions of individuals and groups for whom the pursuit of selfishness reigned supreme. On the other hand, according to Hegel, the state has revealed itself progressively in world history as the one organization predestined to identify the pubic good and private interests and to mediate between them.

If it is to generate fruitful insights, the analytical distinction between state and society needs to be amended in two ways. First, Hegel's analysis reified the state. The "commanding heights" of state power are occupied by concrete actors, individuals belonging to real groups with particular interests and ideologies. The Prussian bureaucracy which Hegel admired so much was manned by members of the East Elbian gentry. The state thus offers to a particular class of people a mechanism of political organization and control which it denies to others. This does not mean that state interests will necessarily express only the particularistic interests of specific classes or status groups. But it makes it probable that in the pursuit of a strategy of foreign economic policy the national interest and the public good will not coincide.

The distinction between state and society connotes a gap between the public and the private sector which exists today in no advanced industrial state. The way in which state and society are actually linked is historically conditioned and that link determines to a large extent whether modern capitalism is atomistic, competitive, organized, or statist. In pointing to the effect which private actors have on state policy, "societal" interpretations of foreign economic policy take two different forms. The democratic explanation traces a direct causal chain from mass preferences (the private and public interests of all members of society) translated via elections into government policy (representing public and private interests of state actors). But in the area of foreign economic policy parties and elections are often less important than interest groups in the formulation and implementation of policy. The interest group model of foreign economic policy traces the infusion of private interests into the definition of public preference and the exercise of public choice. It does so through an analysis which focuses on the interrelationships of social sectors and political organizations. In sum, in both the interest group model and the democratic model, foreign economic policy is seen primarily to reflect society pressures.

But the connection between state and society also runs the opposite way. Public policy can shape private preferences. The locus of decisions in a state-centered model of policy lies in the public realm; in many ways states organize the societies they control. This "statist" interpretation of foreign economic policy discounts mass preferences, political parties, and elections, which are viewed as the effects rather than the causes of government policy. Interest groups are not autonomous agents exerting the pressure which shapes policy but subsidiary agents of the state. These two kinds of interrelationships between state and society are, in reality, always mixed. In the contemporary analysis of foreign economic policy, variants of the former (which reflect the Anglo-Saxon experience) usually are discussed in much greater detail than the latter (which reflect the experience of some countries on the European continent and Japan).

Although most prominent interpretations of state and society agree on the symbiotic relations between the two, they differ on questions of evaluation. Conservatives like Friedman or Hayek argue that the expanding role of government has caused the private sector of advanced industrial states to lose vitality in the twentieth century. Liberals like Galbraith dissent, viewing the new osmosis of state and society as a result of the power of the modern corporation. According to Galbraith's analysis, the economy of advanced industrial states is split between technologically advanced, capital-intensive, and oligopolistic segments and technologically backward, labor-intensive, and competitive sectors. Modern business is big, and big business requires big government. Yet modern governments are too weak to deal effectively with the modern corporation. Neo-Marxists view the osmosis of state and society as the outcome of these two developments. Increasing government intervention in the economy not only reflects the inherent instability of modern capitalism, but also constitutes an attempt to compensate for it.

[Comparative analyses] of a number of advanced industrial states help to put the problems of any one country into a broader perspective. In recent writings on the international political economy, this has not always been the case; categories of analysis were sometimes derived from the particular experience of one country. The French tradition, for example, would lead one to the conclusion that the politicization of the international economy is due to the important role which the state plays in regulating domestic society. Although state and society also mesh closely in Japan, the depoliticization strategy adopted by the Japanese government appears to counteract that conclusion. Similarly, the focus on transgovernmental relations and bureaucratic politics in recent writings on transnational relations partly reflects the fragmentation of the American federal bureaucracy. Yet White House politics and Whitehall politics are qualitatively different. . . .

The actors in society and state influencing the definition of foreign economic policy objectives consist of the major interest groups and political action groups. The former represent the relations of production (including industry, finance, commerce, labor, and agriculture); the latter derive from the structure of political authority (primarily the state bureaucracy and political parties). The governing coalitions of social forces in each of the advanced industrial states find their institutional expression in distinct policy networks which link the public and the private sector in the implementation of foreign economic policy. The notion that coalitions and policy networks are central to the domestic structures defining and implementing policy rests on the assumption that "social life is structured—not exclusively of course, but structured nonetheless—by just those formal institutional mechanisms. To disregard such structures at least implies the belief that social reality is essentially amorphous. This does not mean that institutions work as they are intended to work; it does mean that they have an effect."[1]

Central to the concerns of this volume is the analysis of actors in state and society. In focusing on the governing coalitions which define policy objectives and the institutional organization which conditions policy instruments, we are obviously indebted to both Marxist and Liberal interpretations of politics. However, neither the Marxist emphasis on private ownership of the means of production in capitalist states nor the Liberal focus on the market as a mechanism of distribution in advanced industrial societies is very helpful in explaining the different strategies which advanced industrial states pursue in the international political economy. The national studies in this volume point to an explanation, . . . which steers between sociological types and a general theory of capitalism on the one hand and ideal types and a particularistic description of a number of advanced industrial states on the other. Between generality and specificity, comparative analysis offers comprehension of the unique, historically conditioned components common to the ruling coalitions and policy networks of all advanced industrial states.

This . . . comparative approach is also indebted to the tradition of political research which postulates the state as a central actor in all governing coalitions and a critical institution in all policy networks. The debate between mercantilists on one side and Marxists and Liberals on the other centers on the degree of autonomy of state power from class structure and group conflict. Are the purposes and instruments of state power fundamentally autonomous although modified by the character of a society's class structure and group conflict before they find expression in strategies of foreign economic policy? Or are the purposes and instruments of state power not an independent but merely an intervening variable which aggregates and legitimizes the interests of particular classes or groups? Although I suspect that an answer to these questions does not depend on empirical evidence alone, it is certain that no answers can be tendered in the abstract. Many have understood Marx's *18th Brumaire* as an argument for the relative autonomy of state power. Yet, as a social scientist concerned with questions of empirical evidence, Marx argued for the relative autonomy of the French state in a particular phase of its historical evolution. Speaking generally, this is the kind of answer this volume gives. State power is stronger in some countries, such as Japan or France, than in others; it varies according to whether one analyzes the definition of objectives or the implementation of foreign economic policy.

In their political strategies and domestic structures, the United States, Britain, West Germany, Italy, France, and Japan fall into three distinct groups. In pursuing a liberal international economy as a key objective, policy makers in the two Anglo-Saxon states rely, by and large, on a limited number of policy instruments which affect the entire economy rather than particular sectors or firms. Policy makers in Japan, on the other hand, can pursue their objective of economic growth with a formidable set of policy instruments which impinge on particular sectors of the economy and individual firms. The three continental states occupy an intermediary position with West Germany and Italy showing some affinity for the Anglo-Saxon pattern and France sharing some resemblance to Japan. Corresponding differences also exist in the domestic structures of these six states. In the two Anglo-Saxon countries the coalition between business and the state is relatively unfavorable to state officials and the policy network linking the public with the private sector is relatively fragmented. In Japan, on the other hand, state officials hold a very prominent position in their relations with the business community and the policy network is tightly integrated. The three continental countries hold an intermediary position, with West Germany resembling the Anglo-Saxon pattern, France approximating more the Japanese model and Italy falling somewhere in between.

. . . . The loss of control deplored in the foreign ministries of all advanced industrial states is rooted not only abroad but at home. Lack of action, or inappropriate action, taken in domestic politics often leads to serious consequences in the international political economy. In itself a global approach to meeting global needs appears to be an inefficient and ineffective way of trying to cope with the problems of the international political economy. The management and the analysis of interdependence must start at home.

Note

1. Reinhard Bendix, "Introduction," in Reinhard Bendix, et. al., eds., *State and Society* (Berkeley: University of California Press, 1973), p. 11.

7

Rational Choice Analysis

Rational choice IPE extends the logic and analytic techniques of neoclassical economics to the study of political questions. It is, therefore, rooted in the liberal tradition. Although much rational choice analysis is consistent with its liberal heritage, in recent years it has been employed to bolster some neomercantilist and Marxist arguments. This crossover is but one example of the eclecticism common to contemporary IPE theory. The two excerpts offered here fall within the purview of liberalism, but they provide a good overview of the rudiments of rational choice analysis and the prospects it offers for broader theoretical synthesis.

Mancur Olson and Richard Zeckhauser's classic, "An Economic Theory of Alliances," analyzes issues of international organization, especially the North Atlantic Treaty Organization (NATO). Drawing on Olson's earlier book, *The Logic of Collective Action,* they develop a model that explains why some NATO members bear a disproportionate share of the costs of maintaining the alliance and why NATO goals are not fulfilled optimally. At the heart of their argument is the concept of a "public good," a special situation in which all members of a group benefit irrespective of an individual's contribution to the provision of the good. Security, NATO's primary objective, is a public good because the absence of war benefits all countries of a particular region regardless of the defense burden of individual states. Olson and Zeckhauser contend that the collective action necessary for securing a public good is beset by a paradox because group members seek to maximize their individual benefits and minimize their individual costs. Since each knows that it will enjoy the good even if it does not pay an equal share of the requisite cost, then all will tend to evade their collective duties and, as a consequence, the public good may not be provided. Thus, international organizations are inherently inefficient; the perverse logic of collective action works against the fulfillment of common interests.

Why, then, has NATO not disintegrated? Quite simply, some members place a higher value on the alliance and, therefore, provide a disproportionate share of its public good. For a variety of reasons, the absolute value that the United States places on defense is greater than Belgium's. If they were not drawn together in an alliance, the United States would naturally spend more on defense, even in per capita terms, than Belgium. When the United States and Belgium join together, the military power of the United States is so great that it provides, in an of itself, much of Belgium's defense requirements. This creates a disincentive for Belgium to bear its "fair" share of alliance costs. Olson and Zeckhauser demonstrate with indifference maps how "larger" countries—those that place a higher value on a particular public good—may thus be exploited by smaller countries. In sum, disproportionality is due to neither the "moral superiority" of some states nor the irresponsibility of others; rather, it is the logical outcome of individual states rationally pursuing their own best interests.

Olson and Zeckhauser's argument resonates, to a certain extent, with international relations realism, a precursor of neomercantilist IPE. The efficacy of international organizations, according to realism, is a function of the particular national interests of member states. Insofar as this conclusion is consistent with the logic of collective action outlined above, rational choice analysis and realism are compatible. On the other hand, rational choice analysis does not rest upon the historical specificity embraced by many realist theorists. Instead, it deduces individual behavior from a few core assumptions.

Bruno Frey explicates the postulates of rational choice analysis. The perspective envisions a world of "economic men." Individuals are held to be rational utility maximizers with constant and transitively ordered preferences. The approach focuses on how such individuals strategically interact to secure their interests. It thus assumes that a single logic of formal rationality (as opposed to questions of substantive rationality) motivates a very wide range of human activity, from peasant revolution in Vietnam to nuclear deterrence. Rational choice analysis can be very simple, considering the interactions of only two individuals, or it may be highly complex, examining the strategies of many. Its assumption of rational individuality can be applied across levels of analysis, from individual states in an international system to individual consumers within one country. Frey illustrates how rational choice arguments move well beyond the special case of public goods as evidenced by "politicometric modelling" and game theory. He also notes that the methodology of rational choice is positivist; its abstract logic must be challenged by rigorous empirical scrutiny.

Frey provides examples of rational choice analysis in two areas: tariffs and international organization. On tariffs, rational choice is able to explain why Adam Smith's ideal of free trade is not realized. The benefits of a tariff reduction are a public good; all consumers gain regardless of their effort in bringing it about. Individuals, therefore, have no incentive to incur the "costs" of time and energy required to secure lower tariffs. Conversely, import-competing interests that are threatened by lower tariffs are more willing to assume the organizational costs of lobbying and political action. If the political system is open to specific social pressure, tariffs will not be reduced, because those who stand to gain from lower tariffs are diffuse and unconcerned while potential losers are focused and active. Moreover, Frey argues, the "supply" of tariffs may serve the interests of certain public officials, a situation that further undermines the chances for lower tariffs. The political market tends to favor protection. Regarding international organizations, Frey discusses rational choice analyses that confirm and refine Olson and Zeckhauser's conclusions. In addition, he shows how "constitutional agreements," in the form of various voting rules, might be employed to overcome some of the problems of collective action.

Although he states that rational choice analysis is a variant of economic reasoning, Frey suggests that it opens the way for new syntheses in IPE theory. Some findings of rational choice studies may complement "political science-based" IPE arguments. However, more extensive use of rational choice analysis within theoretical systems not based in liberalism raises an important epistemological question: is the central tenet of individual rationality consistent with the assumptions and argumentation of nonliberal theories? This issue has sparked heated debate among Marxists and is likely to be an ongoing controversy of rational choice analysis.

An Economic Theory of Alliances

MANCUR OLSON, JR. AND RICHARD ZECKHAUSER

I. Introduction

This article outlines a model that attempts to explain the workings of international organizations, and tests this model against the experience of some existing international institutions. Though the model is relevant to any international organization that independent nations establish to further their common interests, this article emphasizes the North Atlantic Treaty Organization, since it involves larger amounts of resources than any other international organization, yet illustrates the model most simply. . . .

There are some important respects in which many observers in the United States and in some other countries are disappointed in NATO and other ventures in international cooperation. For one thing, it is often argued that the United States and some of the other larger members are bearing a disproportionate share of the burden of the common defense of the NATO countries, and it is at least true that the smaller members of NATO devote smaller percentages of their incomes to defense than do larger members. There is also some concern about the fact that the NATO alliance has systematically failed to provide the number of divisions that the NATO nations themselves have proclaimed (rightly or wrongly) are necessary or optimal. . . .

Some suppose that the apparent disproportion in the support for international undertakings is due largely to an alleged American moral superiority, and that the poverty of international organizations is due to a want of responsi-

This article originally appeared in *Review of Economics and Statistics*, Vol. XLVIII, No. 3, August 1966. Reprinted with permission of Review of Economics and Statistics. (c) President and Fellows of Harvard College, 1966.

Table 1. NATO Statistics: An Empirical Test

Country	Gross National Product 1964 (billions of dollars)	Rank	Defense Budget as Percentage of GNP	Rank	GNP Per Capita	Rank
United States	569.03	1	9.0	1	$2,933	1
Germany	88.87	2	5.5	6	1,579	5
United Kingdom	79.46	3	7.0	3	1,471	8
France	73.40	4	6.7	4	1,506	6
Italy	43.63	5	4.1	10	855	11
Canada	38.14	6	4.4	8	1,981	2
Netherlands	15.00	7	4.9	7	1,235	10
Belgium	13.43	8	3.7	12	1,429	9
Denmark	7.73	9	3.3	13	$1,636	3
Turkey	6.69	10	5.8	5	216	14
Norway	5.64	11	3.9	11	1,484	7
Greece	4.31	12	4.2	9	507	12
Portugal	2.88	13	7.7	2	316	13
Luxembourg	.53	14	1.7	14	1,636	4

Ranks:
GNP 1 2 3 4 5 6 7 8 9 10 11 12 13 14
Defense Budget 1 6 3 4 10 8 7 12 13 5 11 9 2 14
as % of GNP
GNP Per Capita 1 5 8 6 11 2 10 9 3 14 7 12 13 4

Source: All data are taken from the Institute for Strategic Studies, *The Military Balance 1965–1966* (London, November 1965).

bility on the part of some other nations. But before resorting to any such explanations, it would seem necessary to ask whether the different sized contributions of different countries could be explained in terms of their national interests. Why would it be in the interest of some countries to contribute a larger proportion of their total resources to group undertakings than other countries? The European members of NATO are much nearer the front line than the United States, and they are less able to defend themselves alone. Thus, it might be supposed that they would have an interest in devoting larger proportions of their resources to NATO than does the United States, rather than the smaller proportions that they actually contribute. And why do the NATO nations fail to provide the level of forces that they have themselves described as appropriate, i.e., in their common interest? These questions cannot be answered without developing a logical explanation of how much a nation acting in its national interest will contribute to an international organization.

Any attempt to develop a theory of international organizations must begin with the purposes or functions of these organizations. One purpose that all such organizations must have is that of serving the *common* interests of member states. In the case of NATO, the proclaimed purpose of the alliance is to protect the member nations from aggression by a common enemy. Deterring aggression against any one of the members is supposed to be in the interest of all. The analogy with a nation-state is obvious. Those goods and services, such as defense, that the government provides in the *common* interest of the citizenry, are usually called "public goods." An organization of states allied for defense similarly produces a public good, only in this case the "public"—the members of the organization—are states rather than individuals.

Indeed, almost all kinds of organizations provide public or collective goods. Individual interests normally can best be served by individual action, but when a group of individuals has some common objective or collective goal, then an organization can be useful. Such a common objective is a collective good, since it has one or both of the following properties: (1) if the common goal is achieved, everyone who shares this goal automatically benefits, or, in other words, nonpurchasers cannot feasibly be kept from consuming the good, and (2) if the good is available to any one person in a group it is or can be made available to the other members of the group at little or no marginal cost. Collective goods are thus the characteristic outputs not only of governments but of organizations in general.

Since the benefits of any action an individual takes to provide a public or organizational good also go to others, individuals acting independently do not have an incentive to provide optimal amounts of such goods. Indeed, when the group interested in a public good is very large, and the share of the total benefit that goes to any single individual is very small, usually no individual has an incentive voluntarily to purchase any of the good, which is why states exact taxes and labor unions demand compulsory membership. When—as in any organization representing a limited number of nation-states—the membership of an organization is relatively small, the individual members may have an incentive to make significant sacrifices to obtain the collective good, but they will tend to provide only suboptimal amounts of this good. There will also be a tendency for the "larger" members—those that place a higher absolute value on the public good—to bear a disproportionate share of the burden, as the model of alliances developed below will show.

II. The Model

When a nation decides how large a military force to provide in an alliance, it must consider the value it places upon collective defense and the other, nondefense, goods that must be sacrificed to obtain additional military forces. The value each nation in an alliance places upon the alliance collective good vis-à-vis other goods can be shown on a simple indifference map, such as is shown in Figure 1. This is an ordinary indifference map cut off at the present income line and turned upside down. Defense capability is measured along the horizontal axis and valued positively. Defense spending is measured along the vertical axis and valued negatively. The cost curves are assumed to be linear for the sake of simplicity. If the nation depicted in Figure 1 were not a part of any alliance, the amount of defense it would obtain (OB) could be found by drawing a cost curve coming out of the origin and finding the point (point A)

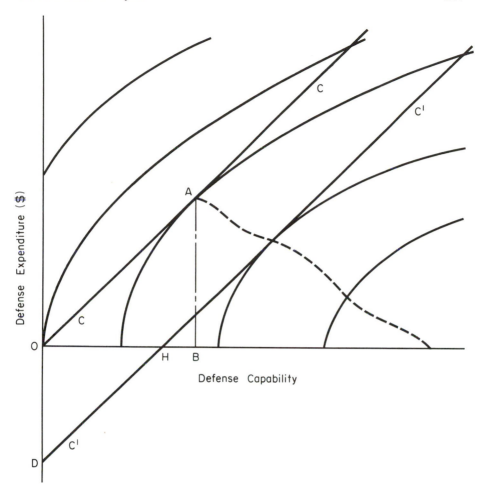

Figure 1. Indifference Map

where this cost curve is tangent to the "highest" (most south-easterly) indifference curve.

In an alliance, the amount a nation spends on defense will be affected by the amount its allies provide. By moving the cost curve down along the vertical axis beneath the origin we can·represent the defense expenditure of allied nations as the distance between the origin and the juncture of the cost curve and the vertical axis. If a nation's allies spend OD on defense, and their cost functions are the same as its own, then it receives OH of defense without cost. This is directly equivalent to an increase in income of OD. The more defense this nation's allies provide, the further the cost constraint moves to the southeast, and the less it spends on defense. By recording all the points of tangency of the total cost curve with the indifference curves, we can obtain this nation's reaction function. The reaction function indicates how much defense this nation will produce for all possible levels of defense expenditure

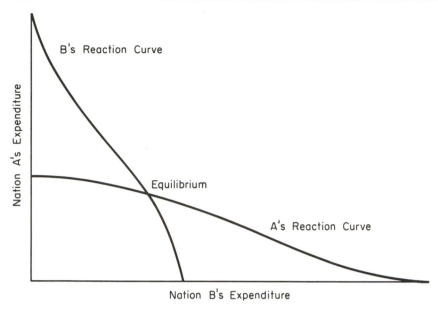

Figure 2. Reaction Curves

by its allies. The amount of defense that this nation provides will in turn influence the defense output of its allies, whose reaction curves can be determined in the same way.

Figure 2 shows the reaction curves for a two-country model (which can easily be generalized to cover N countries). The intersection point of the two reaction curves indicates how much of the alliance good each ally will supply in equilibrium. The two reaction curves need not always intersect. If one nation has a very much larger demand for the alliance good than the other, its reaction curve may lie at every point outside that of the other, in which case it will provide all of the defense. . . .

In equilibrium, the defense expenditures of the two nations are such that the "larger" nation—the one that places the higher absolute value on the alliance good—will bear a *disproportionately* large share of the common burden. It will pay a share of the costs that is larger than its share of the benefits, and thus the distribution of costs will be quite different from that which a system of benefit taxation would bring about. This becomes obvious when income effects—i.e., the influence that the amount of non-defense goods a nation has already forgone has on its desire to provide additional units of defense—are neglected. This is shown in Figure 3 below, which depicts the evaluation curves of two nations for alliance forces. The larger nation, called Big Atlantis, has the higher, steeper valuation curve, V_B, because it places a higher absolute value on defense than Little Atlantis, which has evaluation curve V_L. The CC curve shows the costs of providing defense capability to each nation, since both, by assumption, have the same costs. In isolation, Big

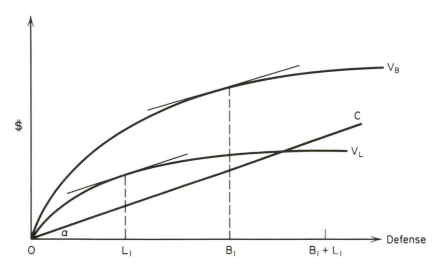

Figure 3. Evaluation Curves

Atlantis would buy B_1 units of defense and Little Atlantis L_1, for at these points their respective valuation curves are parallel to their cost functions. If the two nations continued to provide these outputs in alliance each would enjoy B_1 plus L_1 units of defense. But then each nation values a marginal unit at less than its marginal cost. Big Atlantis will stop reducing its output of deterrence when the sum applied by the two nations together is B_1. When this amount (or any amount greater than L_1) is available, it is not in Little Atlantis' interest to supply any defense whatever. The two nations are therefore simultaneously in equilibrium *only* when Big Atlantis provides B_1 of defense and Little Atlantis provides no defense whatever.

The disproportionality in the sharing of burdens is less extreme when income effects are taken into account, but it is still important. This can be seen most easily by supposing that Big Atlantis and Little Atlantis are identical in every respect save that Big Atlantis is twice the size of Little Atlantis. Per capita incomes and individual tastes are the same in both countries, but the population and GNP of Big Atlantis are twice that of Little Atlantis. Now imagine also that Big Atlantis is providing twice as much alliance defense as Little Atlantis, as proportionality would require. In equilibrium, the marginal rate of substitution of money for the alliance good (MRS) must equal marginal cost for each of these countries, i.e., $MRS_{Big} = MRS_{Little}$ = marginal cost. But (since each country enjoys the same amount of the collective good) the MRS of Big Atlantis is double that of Little Atlantis, and (since the cost of an additional unit of defense is the same for each country) either Big Atlantis will want more defense or Little Atlantis will want less (or both will be true), and the common burden will come to be shared in a disproportionate way.

There is one important special case in which there will be no tendency toward disproportionality. That is when the indifference maps of the member nations are such that any perpendicular from the ordinate would intersect all indifference curves at points of equal slope. In this case, when the nation's cost constraint moves to the right as it gets more free defense, it would not reduce its own expenditure on defense. In other words, none of the increase in income that the nation receives in the form of defense is spent on goods other than defense. Defense in this situation is, strictly speaking, a "superior good," a good such that all of any increase in income is used to buy the good.

This special case may sometimes be very important. During periods of all-out war or exceptional insecurity, it is likely that defense is (or is nearly) a superior good, and in such circumstances alliances will not have any tendency toward disproportionate burden sharing. The amount of allied military capability that Great Britain enjoyed in World War II increased from 1941 to 1944 as the United States mobilized, adding more and more strength to the allied side. But the British war effort was maintained, if not increased, during this period.

Although there is then one exception to the rule that alliance burdens are shared disproportionately, there is no equivalent exception to the rule that alliances provide suboptimal amounts of the collective good. The alliance output will always be suboptimal so long as the members of the alliance place a positive value on additional units of defense. This is because each of the alliance members contributes to the point where its MRS for the good equals the marginal cost of the good. In other words, the result of independent national maximization in an alliance, when the cost function is linear and the same for all members, is that $MRS_1 = MRS_2 = \ldots MRS_N = MC$. There could be an optimal quantity of the collective good only if the total value which all of the alliance members together placed on an additional unit of the good equalled marginal cost, i.e., only if $MRS_1 + MRS_2 + \ldots MRS_N = MC$. The individual nations in an alliance would have an incentive to keep providing additional alliance forces until the Pareto-optimal level is reached only if there were an arrangement such that the alliance members shared marginal costs in the same proportions in which they shared additional benefits (that is, in the same ratio as their marginal rates of substitution of money for the good). When there is such a marginal cost sharing scheme, there need be no tendency toward disproportionality in the sharing of burdens.

III. Qualifications and Elaborations

One simplification assumed in the forgoing model was that the costs of defense were constant to scale and the same for all alliance members. Although military forces are composed of diverse types of equipment and manpower, and thus probably vary less both in cost from one country to another and with scale of output than many single products, it is still unlikely that costs are constant and uniform. For some special types of weapon systems there are undoubtedly striking economies of large scale production, and for conventional ground forces there are probably rising costs as larger proportions of a nation's population are called to arms. Because of this latter tendency, a small country can perhaps get a considerable amount of conventional capability with the first few percentiles of its national income. This tends to keep the military expenditures of small nations in an alliance above the very low level implied by our constant cost assumption. In any event, cross-country variations in marginal costs should not normally alter the basic conclusions deduced from the model. The differences in the amounts which member nations would be willing to pay for marginal units of an alliance good are typically so great that the cost differentials could hardly negate their effect. Even if there were very large differences in marginal costs among nations, there is no reason to assume that national cost functions would vary systematically with the valuation a country places on alliance forces.

A nation's valuation of alliance forces obviously depends not only on its national income, but also on other factors. A nation on the enemy's border may value defense more than one some distance away. A nation that has a large area and long frontiers in relation to its resources may want a larger army than a compact country. On the other hand, if bomb and missile attacks are the main danger, a crowded country may wish to invest more in defense against attack by air. Similarly, a nation's attitudes or ideologies may partly determine its evaluation of defense. Many observers think that the uniformity and intensity of anti-communism is greater among the NATO countries with the highest per capita incomes, and these also happen to be the largest countries in the alliance. It also seems that many people in small and weak countries, both inside and outside of NATO, tend to be attracted to neutralist or pacifist ideologies. This pattern of attitudes may perhaps be partly explained by our model for it suggests that small nations, which find that even large sacrifices on their part have little effect on the global balance, would often be attracted to neutral or passive foreign policies, and that large nations which know that their efforts can decisively influence world events in their own interest will continually need to emphasize the urgency of the struggle in which they are engaged. The popularity of pacific ideologies, the frequent adoption of neutralist policies in small and weak countries, and the activist attitudes and policies of the United States and the Soviet Union are at least consistent with our model.

Whatever the reasons for the different evaluations different nations have for military capabilities in an alliance, the model here still applies. If two countries in an alliance had equal national incomes, but one was more concerned about the common enemy for geographic, ideological, historical, or other reasons, the more concerned nation would not only put a higher valuation on the alliance's military capacity, but would bear a share of the total alliance costs that was even greater than its share of the total benefits. The model deals with the general case of differences in the absolute valuation that nations put upon additional units of an alliance good, whether these differences are due to differences in national income or to other reasons.

Another assumption in the model developed in the foregoing section was that the military forces in an alliance provide only the collective benefit of alliance security, when in fact they also provide purely national, non-collective benefits to the nations that maintain them. When Portugal mobilizes additional forces to suppress the independence movement in Angola, a national goal unrelated to the purposes of NATO, she may at the same time be increasing the total strength of the alliance. Similarly, allied nations may be suspicious of one another, even as they cooperate in the achievement of common purposes, and may enlarge their military forces because of conceivable future conflicts with each other. In any situations in which the military forces of alliance members provide important non-collective benefits as well as alliance benefits, the degree of suboptimality and the importance of the disproportionality will decrease because the non-collective benefits give the member nations an incentive to maintain larger forces.

This fact leads to the paradoxical conclusion that a *decline in the amity, unity, and community of interests among allies need* not *necessarily reduce the effectiveness of an alliance,* because the decline in these alliances "virtues" produces a greater ratio of private to collective benefits. This suggests that alliances troubled by suspicions and disagreements may continue to work reasonably well. To be sure, the degree of coordination among the allies will decline, and this will reduce the efficiency of the alliance forces (in a sense leaving them on a poorer production function), but the alliance forces will be larger.

However important the non-collective benefits of alliances may be, there can be little doubt that above all alliances produce public goods. It is not easy to think of alliances that provide only private goods, though such alliances are perhaps conceivable. If nations simply trade sites for military bases, no common interests or public goods would necessarily be involved. An alliance might also be set up simply to provide insurance in the sense that two nations without any common purpose or common enemy would agree to defend each other in case of attack, but in which neither knew in advance which would suffer aggression. On the other hand, if these two nations thought (as they presumably would) that the fact of their alliance would make it less profitable for other nations to attack either of them, the alliance would provide a public good—a degree of deterrence that could deter an attack on either or both of these nations about as well as it could deter an attack on one alone. There is, moreover, no reason to describe a mere transaction in private goods as an alliance, and the word does not normally appear to be used in that way. A transaction in private goods would be quite as useful between enemies as between "allies," and would normally be completed by a single pair of actions or a single agreement which would not require the continuing consultation, cooperation, and organization characteristic of alliances.

Normally, an additional member can be added to an alliance without substantially subtracting from the amount of defense available to those already in the alliance, and any good that satisfies this criterion is by definition a public good. Suppose two nations of the same size face a common enemy with an army larger than either of them provides by itself. They then form an alliance and maintain a level of military forces larger than either of them had before, but smaller than the sum of their two pre-alliance armies. After alliance both nations enjoy (1) more military security, and (2) lower defense costs, than they had before. This result comes about, not only because a military force can often deter attack by a common enemy against an additional nation without a substantial increase in cost, but also because an alliance may make a greater level of security economically feasible and desirable, and the gains from obtaining this extra security can leave both nations better off.

Another defining characteristic that is sufficient (but not necessary) to distinguish a collective good is that the exclusion of those who do not share the cost of the good is impractical or impossible. Typically, once an alliance treaty has been signed, a member nation is legally bound to remain a member for the duration of the treaty. The decisions about how the common burden is to be shared are not, however, usually specified in the alliance treaty. This procedure works to the disadvantage of the larger countries. Often the smaller and weaker nations gain relatively more from the existence of an alliance than do the larger and stronger powers, and once an alliance treaty has been signed the larger powers are immediately deprived of their strongest bargaining weapon—the threat that they will not help to defend the recalcitrant smaller powers—in any negotiations about the sharing of the common burden. Even at the time an alliance treaty is negotiated exclusion may very well not be feasible, since most alliances are implicit in an already existing danger or goal common to some group of states. That common danger or goal gives the nations that share it an incentive tacitly to treat each other as allies, whether or not they have all signed a formal agreement. A nation can only lose from having another nation with whom it shares a common interest succumb to an enemy, for that will strengthen the enemy's side at the expense of the first nation. It may well be that most alliances are never embodied in any formal agreement. Sometimes a nation may have a geo-political position (e.g., behind an alliance member serving as a buffer state) such that it would be unusually difficult, if not impossible, to deny it the benefits of alliance protection. Then, if it regards alliance membership as a special burden, it may have an incentive to stay out of, or when legally possible to withdraw from, the alliance's formal organization.

This paper also made the simplifying assumption that no alliance member will take into account the reactions other members may have to the size of its alliance contribution. The mutual recognition of oligopolistic interdependence can be profoundly important in small groups of firms, but in the NATO alliance at least, it seems to have been somewhat less important (except with respect to the infrastructure which will be considered later). There are at least two important reasons why strategic bargaining interaction is often less important in alliances than in oligopolistic industries. First, alliances are often involved in situations that contain a strong element of irreversibility. Suppose that the United States were to threaten to cut its defense spending to nothing to get its allies to bear larger shares of the NATO burden. The Soviet Union, if it has the characteristics that American policy assumes, would then deprive the United States of its independence, in which case future defense savings would have little relevance. The United States threat would have only a limited credibility in view of the irreversibility of this process. The second factor which limits strategic bargaining interaction among alliance members stems from an important difference between market and non-market groups. In an oligopolistic group of firms, any firm knows that its competitors would be better off if it were made bankrupt or otherwise driven out of the industry. Large firms thus sometimes engage in price wars or cut-throat competition to drive out the smaller members of an oligopolistic group. By contrast, non-market groups and organizations, such as alliances, usually strive instead for a larger membership, since they provide collective goods the supply of which should increase as the membership increases. Since an ally would typically lose from driving another member out of an alliance, a bargaining threat to that effect may not be credible. This will be especially true if the excluded nation would then fall to the common enemy and (as we argued before) thereby strengthen the enemy at the expense of the alliance.

Even when strategic interaction is important in alliances, the advantage paradoxically still rests in most cases with the smaller nations. There are two reasons for this. First, the large country loses more from withholding an alliance contribution than a small country does, since it values a given amount of alliance force more highly. In other words, it may be deterred by the very importance to itself of its own alliance contribution from carrying out any threat to end that contribution. Second, the large country has relatively less to gain than its small ally from driving a hard bargain. Even if the large nation were successful in the bargaining it would expect only a relatively small addition to the alliance force from the small nation, but when the small nation succeeds in the bargaining it can expect a large addition to the alliance force from the large nation. There is, accordingly, no reason to expect that there is any disparity of bargaining in favor of the larger ally that would negate the tendency toward disproportionality revealed by our model.

IV. Empirical Evidence

When other things are equal, the larger a nation is, the higher its valuation of the output of an alliance. Thus, if our model is correct, the larger members of an alliance should, on the average, devote larger percentages of their national incomes to defense than do the smaller nations. This prediction is tested against the recent data on the NATO nations in Table 1. The following specific hypotheses are used to test the model's predictions:

H_1— In an alliance, there will be a significant positive correlation between the size of the member's national income and the percentage of its national income spent on defense. This hypothesis will be tested against:

H_0— There will not be a significant positive correlation between the variables specified in H_1.

. . . . The Spearman rank correlation coefficient for Gross National Product and *defense budget as a percentage of* GNP is .490. On a one-tailed test this value is significant at the .05 level. We therefore reject the null hypothesis and accept H_1. There is a significant positive correlation indicating that the large nations in NATO bear a disproportionate share of the burden of the common defense. Moreover, this result holds even when the level of per capita income is held constant. [Hypotheses on U.N. funding and foreign aid are also tested and supported.]

Our model indicated that when the members of an organization share the costs of marginal units of an alliance good, just as they share in the benefits of additional units of that good, there is no tendency toward disproportionality or suboptimality. In other words, if each ally pays an appropriate percentage of the cost of any additional units of the alliance good, the results are quite different from when each ally pays the full cost of any amount of the alliance good that he provides. The costs of the NATO infrastructure (common supply depots, pipelines, etc.), unlike the costs of providing the main alliance forces, are shared according to percentages worked out in a negotiated agreement. Since each ally pays some percentage of the cost of any addition to the NATO infrastructure, we have here a marginal cost sharing arrangement.

Thus our model suggests that the burdens of the NATO infrastructure should be borne quite differently from the rest of the NATO burdens. There are other reasons for expecting that the infrastructure burden would be shared in a different way from the main NATO burdens. For one thing, the infrastructure facilities of NATO are all in continental European countries, and ultimately become the property of the host nation. Their construction also brings foreign exchange earnings to these countries, which for the most part are the smaller alliance members. In addition, infrastructure costs are very small in relation to the total burden of common defense, so a small nation may get prestige at a relatively low cost by offering to bear a larger percentage of the infrastructure cost. There are, in short, many private benefits for the smaller alliance members resulting from the infrastructure expenditures. Because of these private benefits, and more important because of the percentage sharing of marginal (and total) costs of the infrastructure, we would predict that the larger members of the alliance would bear a smaller share of the infrastructure burden than of the main alliance burdens.

This prediction suggests that the following hypotheses be tested:

H_4— In an alliance in which the marginal costs of certain activities are *not* shared (but fall instead upon those members who have an incentive to provide additional units of the alliance good by themselves), and in which the marginal costs of other activities are shared (so that each member pays a specified percentage of any costs of these activities), the *ratio* of a member's share of the costs of the activities of the former type to his share of the costs of activities of the latter type will have a significant positive correlation with national income.

H_0— There will be no significant positive correlation between the variables in H_4.

Table 2. NATO Infrastructure

Country	National Income 1960[a] (billions of dollars) (1)	Infrastructure % Reconsidered in 1960[b] (2)	R = (2)/(1) (3)	Military Budget 1960 (billions of dollars) (4)	T = (4)/(2) (5)
United States	411.367	36.98	.0899	41.000	1.1087
Germany	51.268	13.77	.2686	2.072	.1504
United Kingdom	57.361	9.88	.1722	4.466	.4520
France	43.468	11.87	.2731	3.311	.2789
Italy	24.950	5.61	.2248	1.076	.1922
Canada	28.178	6.15	.2183	1,680	.2732
Netherlands	9.246	3.51	.3800	.450	.1282
Belgium	8.946	4.39	.4907	.395	.0900
Turkey	4.929	1.75	.3550	.244	.1394
Denmark	4.762	2.63	.5569	.153	.0582
Norway	3.455	2.19	.6338	.168	.0767
Greece	2.684	.87	.3242	.173	.1989
Portugal	2.083	.28	.1344	.093	.3321
Luxembourg	.386	.17	.4404	.007	.0412

Ranks:														
(1)	1	3	2	4	6	5	7	8	9	10	11	12	13	14
(3)	14	9	12	8	10	11	5	3	6	2	1	7	13	4
(5)	1	8	2	4	7	5	10	11	9	13	12	6	3	14

[a]United Nations, *Yearbook of National Accounts Statistics*, (New York, 1964); and *Balance of Payments Yearbook*, Vol. 15 (Washington, D.C.: International Monetary Fund, 1964).

[b]Charles Croot, "Coordination in the Sixties," reprinted from *NATO Letter* (August 1960).

To test these hypotheses we calculated the correlation coefficient between *national income* and *variable T* in Table 2. The Spearman rank correlation coefficient between these variables is .582, which is significant at the .05 level. We therefore reject the null hypothesis and conclude that the larger members bear a larger proportion of the costs of the main NATO forces than they do of those NATO activities for which the costs of each unit are shared. The difference between the distribution of infrastructure costs and the distribution of alliance burdens generally is quite striking, as the tests of the following hypotheses indicate:

H_5— In the NATO alliance there is a significant negative correlation between national income and the percentage of national income devoted to infrastructure expenses.

H_0— There is no significant negative correlation between the variables in H_5.

The Spearman rank correlation coefficient between *national income* and *variable R* in table 2 is $-.538$, which is significant at the .05 level. Thus, not only is it the case that the larger nations pay a smaller share of the infrastructure costs than of other alliance costs; it is also true that there is a significant negative correlation between national income and the percentage of national income devoted to the NATO infrastructure, which is in vivid contrast to the positive correlation that prevails for other NATO burdens. This confirms the prediction that when there are marginal cost sharing arrangements, there need no longer be any tendency for the larger nations to bear disproportionately large shares of the costs of international organizations. If it happens at the same time that the smaller nations get greater than average private benefits from their contributions, they may even contribute greater percentages of their national incomes than the larger members.

V. Conclusions and Recommendations

All of the empirical evidence tended to confirm the model. . . . In NATO there is . . . a statistically significant positive correlation between the size of a member's national income and the percentage of its national income devoted to the common defense.

As our model indicated, this is in part because each ally gets only a fraction of the benefits of any collective good that is provided, but each pays the full cost of any additional amounts of the collective good. This means that individual members of an alliance or international organization have an incentive to stop providing the collective good long before the Pareto-optimal output for the group has been provided. This is particularly true of the smaller members, who get smaller shares of the total benefits accruing from the good, and who find that they have little or no incentive to provide additional amounts of the collective good once the larger members have provided the amounts they want for themselves, with the result that the burdens are shared in a disproportionate way. The model indicated two special types of situations in which there need be no such tendency toward disproportionality. First, in cases of all-out war or extreme insecurity defense may be what was strictly defined as a "superior good," in which case a nation's output of a collective good will not be reduced when it receives more of this good from an ally. Second, institutional arrangements such that the members of an organization share marginal costs, just as they share the benefits of each unit of the good, tend to work against disproportionality in burden sharing, and it is a necessary condition of an efficient, Pareto-optimal output that the marginal costs be shared in the same proportions as the benefits of additional units. The NATO nations determine through negotiation what percentages of any infrastructure expenditure each member will pay, and this sharing of marginal costs has led the smaller members to bear a very much larger share of the infrastructure burden than they do of the other NATO burdens. The fact that the model predicts not only the distribution of the principal NATO burdens, but also the greatly different distribution of infrastructure costs, suggests that the results are in fact due to the processes described in the model, rather than to some other cause.

The model's implication that large nations tend to bear disproportionate shares of the burdens of international organization, and the empirical evidence tending to confirm the model, does *not* entail the conclusion that the small nations should be told they "ought" to bear a larger share of the common burdens. No moral conclusions can follow solely from any purely logical model of the kind developed here. Indeed, our analysis suggests that moral suasion is inappropriate, since the different levels of contribution are not due to different moral attitudes, and ineffective, since the less than proportionate contributions of the smaller nations are securely grounded in their national interests (just as the disproportionately large contributions of the larger countries are solidly grounded in their national interests). Thus, American attempts to persuade other nations to bear "fair" shares of the burdens of common ventures are likely to be divisive and harmful even to American interests in the long run.

The model developed here suggests that the problems of disproportionality and suboptimality in international organizations should be met instead through institutional changes that alter the pattern of incentives. Since suboptimal provision is typical of international organizations, it is possible to design policy changes that would leave everyone better off, and which accordingly may have some chance of adoption. Appropriate marginal cost sharing schemes, such as are now used to finance the NATO infrastructure, could solve the problem of suboptimality in international organizations, and might also reduce the degree of disproportionality. Substituting a union for an alliance or international organization would also tend to bring about optimality, for then the unified system as a whole has an incentive to behave in an optimal fashion, and the various parts of the union can be required to contribute the amounts their common interest requires. Even a union of smaller members of NATO, for example, could be helpful, and be in the interest of the United States. Such a union would give the people involved an incentive to contribute more toward the goals they shared with their then more nearly equal partners. Whatever the disadvantages on other grounds of these policy possibilities, they at least have the merit that they help to make the national interests of individual nations more nearly compatible with the efficient attainment of the goals which groups of nations hold in common.

A final implication of our model is that alliances and international organizations, as presently organized, will not work efficiently, or according to any common conception of fairness, however complete the agreement and community of interest among the members. Though there is obviously a point beyond which dissension and divergent purposes will ruin any organization, it is also true that some differences of purpose may improve the working of an alliance, because they increase the private, non-collective benefits from the national contributions to the alliance, and this alleviates the suboptimality and disproportionality. How much smaller would the military forces of the small members of NATO be if they did not have their private fears and quarrels? . . . The United States, at least, should perhaps not hope for too much unity in common ventures with other nations. It might prove extremely expensive.

The Public Choice View
of International Political Economy

BRUNO S. FREY

There can be no question that the study of international political economy has received insufficient attention in both economics and political science. As Joan Spero puts it,

> in the twentieth century the study of international political economy has been neglected. Politics and economics have been divorced from each other and isolated in analysis and theory. . . . Consequently, international political economy has been fragmented into international politics and international economics.[1]

Though this gap still exists today, it has been narrowed considerably by the emergence of a new field from international relations theory, a field commonly known as "international political economy." . . .

Political science-based scholars quite outspokenly claim international political economy as their proper and exclusive domain. R.J. Barry Jones, for example, argues that "the foundations of a realistic study of the international political economy are not dissimilar to those of conventional political analysis."[2] Accordingly, "power" and "authority" are taken to be the central concepts with which to study problems. In addition, the analysis has to be "dynamic" and has to take historical processes into account. It is not surprising that, as a consequence, there is a marked tendency to reject any approaches based on economic theory. . . .

This rejection of economic theory does not, however, seem to be based on an extensive knowledge of the literature. In particular, the economic approach to politics, usually called "public choice," seems to be almost totally

Reprinted from *International Organization,* Vol. XXXVIII, 1984, Bruno Frey, "The Public Choice View of International Political Economy" by permission of The MIT Press, Cambridge, MA.

disregarded. The "classic" writers in the public choice tradition, such as Arrow, Downs, Buchanan, Tullock, and Niskanen, are hardly, if ever, mentioned. It is due to this oversight that Roger Tooze, writing as recently as 1981, can maintain that "neo-classical theory . . . treats political and social processes perfunctorily, as extraneous, and, at best, exogenous factors."[3] Yet what public choice has done is exactly to treat political processes as *endogenous* factors.

In this article, I endeavor to show that, first, public choice has been applied specifically to international political economy, and that there is a large and rapidly growing literature on the subject. Second, public choice offers an interesting and worthwhile approach to the area, an approach that complements the political science-based views of international political economy in a useful way. Consequently, the claim for exclusivity made by some writers based in political science should be replaced by the realization of the need for mutual cross-fertilization of the two (partly competing) approaches. Section 1 provides a short survey of those parts of the public choice approach most relevant for international political economy.

1. The Public Choice Approach

General Characteristics

Public choice, sometimes called the "economic theory of politics" or "(new) political economy," seeks to analyze political processes and the interaction between the economy and the polity by using the tools of modern (neoclassical) analysis. It provides, on the one hand, an explicit positive approach to the workings of political institutions and to the behavior of governments, parties, voters, interest groups, and (public) bureaucracies; and it seeks, on the other, normatively to establish the most desirable and effective political institutions. Public choice is part of a movement that endeavors to apply the "rational behavior" approach to areas beyond traditional economics. In recent years, an increasing number of political scientists, sociologists, and social psychologists have taken up this approach. It thus constitutes one of the rare successful examples of interdisciplinary research.

Both the rational behavior approach to social problems and public choice theory are characterized by three major features. First, the individual is the basic unit of analysis. The individual is assumed to be "rational" in the sense of responding in a systematic and hence predictable way to incentives: courses of action are chosen that yield the highest net benefits according to the individual's own utility function. Contrary to what nonspecialists often believe, it is not assumed that individuals are fully informed. Rather, the amount of information sought is the result of an (often implicit) cost-benefit calculus. Indeed, in the political arena it often does not pay the individual to be well-informed—this is known as "rational ignorance."

Second, the individual's behavior is explained by concentrating on the changes in the constraints to which he or she is exposed; that is, the preferences are assumed to be constant. Individuals are assumed to be capable of comparing alternatives, of seeing substitution possibilities, and of making marginal adjustments.

The third characteristic is that the analysis stresses rigor (and is sometimes formal). The results must yield a proposition that (at least in principle) can be subjected to econometric or politicometric testing.

There is no need to go into general public choice theory here; only its applications to problems of international political economy are relevant to this discussion.

The Concepts Applied in International Political Economy

In the international field, some theoretical concepts developed in public choice are used particularly often. Public goods theory and politico-economic modeling will be briefly mentioned here in order to illuminate the public choice approach to international political economy.

Public good theory. Public goods is the concept most frequently used within economic-based international political economy. Its usefulness is well illustrated in a contribution by Charles Kindleberger, in which he looks at various aspects of the international economy from the point of view of public goods, and at the tendency for free riding, in which a public good is available to all irrespective of whether they have contributed to its supply.[4] Thus, law and order can be considered a public good forming an important complement to foreign trade. Its absence can lead to serious disruption in international exchange. The institution of the state may also be regarded as a public good. The high costs arising when it does not exist are illustrated by the example of Germany in 1790. At that time there were 1,700 tariff boundaries with three hundred rulers levying tolls as they pleased. Under such circumstances it was no wonder that the advantages of trade exchange could not be exploited to any great degree. The existence of national monetary institutions may also be looked upon as a public good.

There are a great many other applications of the public goods concept and the concomitant free-rider problem, such as trade liberalization, nationalism, alliances, and burden sharing. A further application is the preservation of nature beyond natural frontiers, such as the campaign against whaling or the protection of the atmosphere.

The public goods concept is extremely useful and intuitively plausible. The ease of application may, however, sometimes hide underlying problems. The exact conditions under which free riding occurs are still unknown; often it is simply assumed that actors do not contribute to the common cause. Laboratory experiments on public goods situations suggest that free riding does not occur as often as pure economic theory would have us think. Moreover, institutional conditions are often such that free riding is discouraged.

Even when national actors fully perceive that it is advantageous for them to cooperate in the provision of a public good, it is difficult and sometimes even impossible to coordinate joint action. In view of the general impossibility of forcing independent national actors to cooperate, the free-rider problem can be overcome by finding *rules* or *constitutional agreements* that lay down the conditions for cooperation.

In order to find a set of rules that the participants are willing to accept in a state of (partial) uncertainty about the future (i.e., beyond the veil of ignorance), the actors must believe that obeying the rules will be advantageous to them. The agreement must lead to a beneficial change according to the expectations of all actors (Pareto-superiority), because only under these conditions will there be voluntary cooperation—that is, unanimity among the participants. These conditions are not easily set up and maintained in the international system. Once a set of rules or a constitution has been agreed upon, the problem is to ensure that the rules are observed and that individual nations have no incentive to back out of or attempt to alter the agreement. The "constitutional" approach has been applied to various problems in international political economy, among them environmental and fisheries pacts, international public health accords, cooperation about forecasting (and in the future possible influencing) the weather, the use of outer space, and the international judicial system.

The establishment and enforcement of rules has occupied a central position in two areas. First *international monetary arrangements* may be considered to be, if well designed, advantageous to all, but the incentives for deviation are also marked. It is therefore necessary to consider not only the Pareto-superiority of an international monetary scheme but also the benefits and costs to the individual participating nations. This aspect has been overlooked in the many proposals made in this area; they usually assume (implicitly) that there is a "benevolent international dictator" who will put them into effect.

An important related question is why certain rules have not influenced behavior as much as one might have expected. One example is provided by the Bretton Woods system, in which changes in exchange rates have been made too infrequently, and generally too late. The reason is that forces militate against both devaluation and revaluation. Voters, it is believed, interpret devaluation as an admission of financial failure, with negative consequences for the government in power. Revaluation is good for the voters (as consumers) but very bad for well-organized groups of exporters and import competitors; so the government may again run into trouble. In view of this unwillingness to adjust exchange rates, an agreement allowing freely flexible exchange rates may be preferable because the issue is then taken out of government (and central bank) politics.

The second international area in which rules play an important role is that of *international common property resources*. The need for international conventions and rules is obvious in view of the pollution of the atmosphere and the overfishing and overexploitation of the oceans. The difficulty in reaching agreement on what these rules should be is equally well known. It is hard to obtain consensus because no country can be forced to accept rules. The only acceptable rules are those that produce such high aggregate net benefits that they can be distributed among the participating countries in such a way that everyone finds it advantageous to agree and to stick to the rules. Such rules do not usually exist; it is quite possible that agreement on some of the current proposals concerning common international property resources would be worse for participating countries than no agreement at all.

Politico-economic modeling. Politico-economic models or, as they are often called, political business-cycle models, study the interdependence between the economy and the polity by explicitly analyzing the behavior of actors. They test the resulting propositions using econometric, or rather politico-metric, techniques. The simplest such model analyzes a circular system: the state of the economy influences the voters' evaluation of the government's performance, which is reflected by a vote or government popularity function. If the government considers its chances of re-election to be poor, it uses economic policy instruments to influence the state of the economy and thus the voters' decisions. (The government's actions may depend on its ideology if it considers its re-election changes to be good.) The model is, of course, a great simplification of reality, but it has already been shown that the framework can be extended to incorporate additional actors and relationships.

A politico-economic model for a closed country can be extended in two ways to include international politico-economic relationships. The first approach concentrates on the *internal* connections between the economy and the polity but also introduces international influences. In this case the politico-economic model outlined above is amended by factors emerging from the international sphere. . . .

International political events may also affect votes and government popularity. Empirical studies of the United States show that, when the country is subjected to an international political crisis, the population tends to "rally round the flag." Another influence that may be introduced into politico-economic models is the foreign intervention in a country's internal polity that may occur if a foreign nation considers the results of a particular election undesirable. Government politicians may also have specific international political preferences and influence the internal economy accordingly, provided that their re-election chances are not thereby seriously diminished.

. . . . [T]he use of economic policy instruments is influenced by international economic conditions. The possibility of creating a political business cycle aimed at improving re-election chances depends on institutional conditions with the international economy. An expansionary economic policy yields more favorable short-run inflation-unemployment (or real income) trade-offs within a system of adjustable pegs than with a depreciating exchange rate. A system of adjustable pegs may thus be expected to increase the government's incentive to attempt to gain votes by introducing an expansionary policy before elections and devaluing thereafter.

The second approach goes one step further, by considering the *mutual interdependence* of domestic and foreign economies and polities. This research strategy is particularly well-developed with regard to arms race models. Such models have traditionally analyzed the mutual responses of two nations to each other's defense outlays in a rather mechanistic way. . . . In the last few years, however, the decision-making structure has been greatly improved by the introduction of elements of public choice. In particular, it has been recognized that a nation's response to another nation's armament depends on the government's utility, and is subject to the constraints imposed by the desire to be re-elected as well as by economic resources. The models have been econometrically estimated and their behavior has been analyzed with the help of extensive simulations.

Both of the aforementioned approaches are useful; the second is, of course, much more far-reaching and may therefore be difficult to apply to politico-economic interaction as a whole. It may therefore be advisable to restrict it to one particular issue at a time.

The next two sections illustrate the substantive contributions that public choice has made to the theoretical and empirical analysis of international political economy. No complete survey of all the applications of public choice is intended. I omit, for example, public choice's applications to the interdependence between foreign trade flows and political conditions, and the determinants of foreign direct investment and of international aid. Rather, I concentrate on two particular areas, the formation of tariffs and trade restrictions (section 2) and international organizations (section 3).

2. Tariffs and Trade Restrictions

Most economists approach the analysis of tariffs and other restrictions on trade from the same standpoint. They start from the basic proposition of international trade theory: that free trade leads to higher real income and is desirable not only for the world as a whole but also for individual countries. The problem for political economists of the neoclassical public choice orientation, therefore, is to explain why tariffs nonetheless exist, and why governments so rarely seem to take the welfare-increasing (Pareto-optimal) step of abolishing tariffs. A government might be expected to win votes by abolishing tariffs, either because a majority of the electorate would directly benefit or because the government could redistribute the gains so that a majority of the electorate would be better off than in a situation with tariffs. If citizens were to determine tariffs by a single direct majority vote in an assembly, the *median voter* would vote in favor of free trade.

The simplistic assumptions of the median-voter model must, however, be modified in a number of important respects if it is to represent reality. This provides an explanation for the continuing existence and possibly even growth of tariffs. At least five modifications must be considered.

The first is that the losers in any tariff reduction, the people engaged in the domestic production of the goods concerned, are not compensated. If they form a majority, they will obstruct the reduction or the elimination of tariffs.

The second necessary modification is to consider the fact that prospective gainers have less incentive than prospective voters to participate in the vote, to inform themselves, and to organize and support a pressure group. Tariff reductions are a public good whose benefits are received by everybody, including those not taking the trouble and incurring the cost to bring about the reduction. The prospective cost of tariff reduction to the losers is, however, much more direct and concentrated, so that it is worth their while to engage in a political fight against tariff reduction. In addition, the well-defined short-term losses to be experienced by the losers are much more visible, and therefore better perceived, than uncertain gains to be made in the distant future by the winners. The fight over trade restrictions benefits those sectors protected from competition but otherwise serves no socially useful purpose, because it wastes scarce resources. This aspect is the subject of the *theory of rent seeking*. It is useful to differentiate between two activities, both of which, from society's point of view, waste resources. "Rent seeking" is the activity by which trade restrictions (tariffs, quotas) generate rents to one's advantage; "revenue seeking" is the fight over the distribution of revenues, and is thus a general distributional phenomenon.

A third modification of the simple median-voter model considers the possibility that the prospective losers in a free-trade regime may be better represented in parliament and in the government than the prospective winners, depending on the system of voting.

A fourth modification reflects the fact that logrolling or vote trading can make it possible for two measures, each of which would increase the country's welfare, both to be defeated by a majority. Vote trading may happen if groups of voters have unequal preference intensities for two issues. This is very likely to be the case where tariffs are concerned. Consider a group I of voters engaged in domestic import-competing activities. The group's main preference is against the reduction of tariffs for its *own* products; it weakly favors the reduction of tariffs for some other products. Assume a group II of voters whose main interest lies in maintaining the tariff for the latter products, and who have a weak preference for a tariff reduction for the former products. If neither of the two groups commands a majority, and the other voters perceive the benefits of free trade, the tariff reduction would pass and free trade would be established. If, however, groups I and II combined have a majority, they can agree to trade votes: group I votes against the tariff reduction that group II strongly opposes provided group II votes against the tariff reduction that group I strongly opposes. This leads to a majority vote against tariff reductions, reducing overall welfare. Vote trading, however, does not always decrease social welfare. The fact that a market is established in which mutually advantageous trades can take place may increase social welfare, provided all the actors affected by the relevant activity take part in the vote-trading process.

The final modification of the median-voter model recognizes that tariffs provide revenue for governments, which in their absence would find it even more difficult to finance public expenditure. This is especially true in developing countries where, due to the inefficiency of the tax system, there is little tax revenue. A government will therefore wish to secure this income source and hence will oppose free trade.

These five modifications of the simple median-voter model combine to explain why free trade, which is optimal from the point of view of the country as a whole, is not found in reality. The discussion suggests that there is, on the contrary, a *political market for protection*. Protection is demanded by particular groups of voters, firms, and associated interest groups and parties, and supplied by politicians and public bureaucrats. Thus, the public choice approach to international political economy stresses the importance of interest groups. . . .

Another actor plays an important role in tariff formation: the *public administration*. This body has considerable influence on the "supply side" of tariff setting because it prepares, formulates, and implements trade bills once government and parliament have made a decision.

The activity of public bureaucrats with respect to tariffs may be analyzed with the help of the "rational" model of behavior, for example, by maximizing utility subject to constraints. The main elements in the bureaucrats' utility function may be assumed to be the prestige, power, and influence that they enjoy relative to the group of people they are officially designed to "serve," their clientele. In most cases this clientele will be located in a specific economic sector; in the case of public officials in the ministry of agriculture, for example, the clientele would be those groups with agricultural interests. They are, moreover, proud of being able to show that they are competent to perform their job ("performance excellence"). Public bureaucrats will therefore tend to fight for the interests of "their" economic sector, and will work for tariffs and other import restrictions in order to protect it from outside competition. They will prefer to use instruments under their own control rather than to follow general rules imposed by formal laws. They will thus prefer various kinds of nontariff protection and support (subsidies) to general tariffs.

The public bureaucracy faces constraints imposed by parliament and government. However, both of these actors have little incentive to control public administration more tightly, because they depend on it to attain their own goals. In addition, political actors have much less information available to them than the public bureaucracy does, in particular with respect to the sometimes very complex issues of protection. The limited incentive of politicians to control the public administration gives bureaucrats considerable discretionary power, which they use to their own advantage.

Public choice theory has also been used to try to explain differences in international protection, that is, the intersectoral *structure of tariffs*. It is hypothesized that the more concentrated industries find it easier to organize and to muster political pressure because a smaller number of enterprises is more willing to bear the transaction, organization, and lobbying costs involved in getting tariff protection. . . .

This discussion of the various factors that may be used to explain tariffs and other trade restrictions shows that the study of international political economy based on public choice is well under way, and that useful theoretical and empirical results have been achieved using an approach that differs strongly from political scientists' international political economy. Research is, however, only at an early stage, and various aspects of the analysis must be improved. First, the behavior of the actors (government, interest groups, and public bureaucracy) must be modeled more carefully, taking their characteristic preferences and constraints into account. Second, the equations used for econometric estimation need to be more closely and consistently linked with the theoretical models. Third, the framework of the analysis should be extended, so that all the relevant causal relationships can be included in the analysis. Politico-economic conditions do not only affect tariffs; tariffs also affect the state of the economy and polity. Thus, both directions of the interdependence between tariffs and the political economy should be considered.

3. International Organizations

Interesting contributions have been made within public choice to the study of the benefits and costs of joining international organizations, their decision rules, and their internal bureaucracy, as well as to the study of bargaining in an international setting.

Benefits and Costs of Joining International Organizations

An international organization may perform various services: it may provide public goods and services, coordinate the activities of actors in the international system, and form an institutional setting for alliances. International organizations may also be used to further private (i.e., national) aims; it would therefore be a mistake to assume that they maximize the collective economic welfare either of the individuals of a particular country or of the world as a whole.

Much of the output of international organizations has the character of a public good, thus providing an incentive for countries to behave as free riders. Under these circumstances the organization will only be able to operate effectively if it involves a small group of countries, permitting direct interaction and imposing high costs on free riders; if private goods are offered selectively to the members of the organization, providing an incentive for individual countries to join and participate in the financing of the organization; or if participation is achieved by coercion.

It has been empirically shown that small, regional or international organizations are indeed more successful than large ones. The second option, creating selective incentives for members, is very common in international organizations. The existence of such private goods is a very important bargaining tool used by governments in persuading parliaments to agree to join. Considerable effort is therefore devoted to transforming public goods into private goods owned by the organization. Finally, coercion is difficult and often impossible in an international context, because the member countries are unwilling to give up their independence. However, as long as the international system is composed of sovereign states, the assumption that coercion is possible solves the problem of international organization by definition.

An organization may also be formed if the potential participants' perception of the advantages of membership and the social pressure to belong to it can be increased by education and propaganda. As in the case of coercion, this approach has very little chance of success in the international system. The solution of international politico-economic problems is therefore not to be found in one or several supranational authorities.

The Bureaucracy of International Organizations

It has been suggested that the characteristics of bureaucracies are more pronounced in the international than in the national setting. The main reason is that they have greater room for discretionary action, because there is neither the opportunity nor the incentive to control them. Control is difficult because the "output" of an international organization is undefined and usually cannot be measured. There are no political institutions that would gain by tightly controlling an international organization: national governments would only run into trouble with other national governments if they tried to interfere with the workings of such institutions. They therefore prefer to let things go and only intervene if they feel that their own nationals employed in the organization are being unfairly treated or that their national interests are being threatened by the organization's activity. The lack of incentives is another example of the free-riding problem.

Due to the lack of effective control in an international organization, none of the layers in the hierarchy has any real incentive to work toward the "official product," because their utility depends hardly at all on their contribution. The national quotas for positions that are a feature of many international organizations drive a further wedge between the individual's utility and the organization's official function. This particular incentive structure leads to a growth of international bureaucracies quite independent of the tasks to be performed, because all bureaucrats benefit from larger budgets and a greater number of employees. International bureaucracies are also characterized by a low degree of efficiency and a profusion of red tape, because the formalized internal workings of the organization become dominant. A considerable share of the budget will be used for internal purposes, and to provide side benefits for the bureaucrats themselves.

This theory of international bureaucracy still has to undergo empirical testing.

International Bargaining

Modeling international bargaining is a formidable task, because the process has little structure and involves many variables. There have therefore been few public choice studies of international bargaining. Those that do exist have each concentrated on a particular aspect of the problem.

In international negotiations, linkages between various issues are quite a common feature. R.D. Tillison and T.D. Willett have shown that linkages are more important when the distribution of benefits from agreements is highly biased toward a small number of countries. The linkage of issues whose distributional consequences offset one another can help promote agreements that would otherwise fail because of distributional effects. On the other hand, linkages play a small role when the benefits from an agreement are considered to be "fairly" distributed across countries. In this case a consensus can be reached without introducing an additional dimension in the form of linked issues. Such hypotheses are plausible but, again, have not yet been empirically tested.

4. Concluding Remarks

The aim of this survey has been to show that public choice has made considerable and valuable contributions to international political economy. The approach has shed some new light on the field and should be of interest to all social scientists concerned with bridging the gap between international economics and international (political) relations. Due to limitations of space, I have only been able to provide some characteristic examples, and some selected fields of application, of the public choice approach to international political economy.

The public choice approach to international political economy has both strengths and weaknesses, a proposition, of course, true for any approach, including that adopted by political scientists. Five points merit amplification.

First, the public choice view provides *fresh insights* into the area, in the same way that the economics-based approach illuminated general politics. This is not to claim that the approach is superior to any other; rather, that it is able to illuminate particular aspects of international political economy (while being unable to contribute much in other areas). As will become clear from these concluding remarks, the specific strengths of the public choice view are also responsible for its specific weaknesses. This is true whenever one considers the advantages of applying a new method to an already established field, such as international political economy. There is a tendency to use theoretical and empirical methods without paying sufficient attention to the particular historical and institutional conditions existing in the field of study. A quick application is tempting, because it is seemingly easy to undertake, and the shortcomings of the analysis may not be obvious. It is necessary, however, to investigate thoroughly whether particular theoretical concepts such as public goods and free riding really capture the essential features of reality.

Second, an advantage of the public choice approach to international political economy is that the analysis is based on an explicit and unified *theory of human behavior,* and on a *technical apparatus* capable of producing theoretical solutions and empirically testable propositions. This technical elegance leads, however, to a tendency to sacrifice relevance for rigor. There are already some areas of this type of international political economy where the heavily formalistic apparatus used is out of all proportion to the resulting advances in knowledge.

Third, the public choice approach concentrates on specific aspects of international political economy, making it possible to isolate and analyze relatively simple relationships. The *high degree of abstraction* allows public choice scholars to gain major insights into complex problem areas. But it also involves the danger of leaving out relevant aspects or of keeping constant (by the "ceteris paribus" assumption) variables that are so closely and importantly connected with the problem being studied that they should be considered an endogenous part of the model. While this survey has concentrated on microanalytical and partial analyses, there are approaches within public choice that attempt to provide view (in particular the politico-economic models).

Fourth, the emphasis on deriving propositions that are at least in principle amenable to *empirical testing* is healthy, because it forces reality on the researcher. Econometric, or rather politicometric, analyses also provide factual knowledge about the relationships among the variables being studied. The disadvantage of this empirical orientation is that aspects difficult or impossible to measure quantitatively are easily excluded. Thus, the relationships for which data are easily available are those that tend to be studied. A common shortcoming of empirical economic research is that the operationalization of individual theories is often done in a rather cavalier way. In that respect economists could certainly learn from quantitative political scientists, as well as from other social scientists.

Empirical research has so far been predominantly concerned with the United States. This makes it more difficult to evaluate the contribution of public choice to international political economy, because it is difficult to know what part of the results is due to the public choice view, what part to the particular conditions obtaining in the United States. It is therefore important that empirical tests of the theories should be undertaken for other nations.

Fifth and finally, the public choice view is *interdisciplinary,* in a specific sense of the word: it combines the economic and political aspects of international political economy but uses a single theoretical approach. (Usually, interdisciplinary is understood to mean that theoretical approaches have to be combined.) This has the advantage that the two areas can be fused together, but it carries the already mentioned danger that only selected aspects of their interrelationship will be treated. There can be little doubt, however, that economists engaged in research on international political economy can gain from the work done by political scientists, especially from their experience of the institutions and political processes encountered in the international sphere. Up to now, there has been relatively little contact between public choice researchers and other scholars in the field. This survey will have achieved its goal if it has convinced the reader that the opposite proposition is also true: political scientists would benefit from considering and studying the public choice approach to international political economy.

Notes

1. J.E. Spero, *The Politics of International Economic Relations* (London: Allen & Unwin, 1977), pp. 1–2.

2. R.J. Barry Jones, "International Political Economy: Problems and Issues," Part II, *Review of International Studies* 8 (1982), p. 50.

3. R. Tooze, "Economics, International Political Economy and Change in the International System," in B. Buzan and R.J.B. Jones, eds., *Change and the Study of International Relations: The Evaded Dimension* (London: Pinter, 1981), p. 130.

4. C.P. Kindleberger, *Government and International Trade,* Princeton Essays in International Finance no. 129 (Princeton, N.J., July 1978).

Hegemonic Power and Stability

All three major schools of IPE thought discuss hegemonic international power and its consequences. Unlike rational choice analysis, however, the idea of hegemony is not grounded in a particular paradigm. Consideration of hegemony does not necessarily entail acceptance of certain theoretical assumptions. It is, rather, an empirical question that is examined from different perspectives. Studies of hegemonic power do, nonetheless, open a new avenue for theoretical synthesis. Although each theory qualifies its understanding in peculiar ways, the shared interest in hegemony has drawn the disparate approaches together in a complementary, albeit eclectic, manner.

Immanuel Wallerstein develops a world-systems interpretation of hegemony, derived largely from his reading of the Marxist IPE canon. For Wallerstein, hegemony is a temporary characteristic of the world-system as a whole. It exists when a single country dominates simultaneously global processes of capital accumulation and international balances of military power. This general definition is basically consistent with neorealist and liberal notions of hegemony. Wallerstein differs from these contending theories, however, by emphasizing a particular economic logic of hegemonic power. Although he recognizes the political aspect of the world-system, he focuses on secular trends of economic growth ("A phases") and recession ("B phases") as the driving force of hegemonic rise and decline. These "long cycles," or "Kondratieff cycles" (named for a pioneering Soviet economist) reflect the dynamism of capitalism as a world-economy. Comparative advantages change as the locus of accumulation shifts around the globe. Wallerstein suggests a historical sequence of a country first acquiring dominance in agro-industrial production, then international commerce, and finally world finance. Only when one state enjoys predominance in all three of these areas does the system possess a hegemonic concentration of power.

Hegemony, as such, is very rare. Wallerstein contends that only three countries have attained the status of global hegemon since the seventeenth century: the United Provinces of the Netherlands in the mid-seventeenth century; the United Kingdom in the mid-nineteenth; and the United States in the mid-twentieth. Hegemony is, in addition, very unstable. The systemic forces that create it also work against its durability.

Wallerstein also considers the politics of hegemony. Interstate military balances, he suggests, parallel economic transformations. As hegemonic powers achieve economic dominance, they also gain a clear edge in naval and air power. Wallerstein does not posit a crudely economistic explanation of international military power. Rather, he believes that global economics and politics are interactive and that the "logistics" that explain this interaction have yet to be fully analyzed. Moreover, international politics can, at times, be relatively autonomous from world-economic conditions. Hegemony is not empire; even though economic power can be highly concentrated, political power is distributed in a bipolar or multipolar fashion. The major wars engendered by political decentralization serve to restructure the world-system and usher in periods of hegemony. On the other hand, Wallerstein argues that global economic preeminence clearly shapes the definition of the hegemon's national interest. Simply put, hegemons tend to identify their interests with liberal international trade, investment, and finance.

Ironically, liberalism, which serves the interests of a hegemon at the height of its power, contributes to hegemonic decline. The coincidence of comparative advantages that defines hegemony is undermined by international openness and technological diffusion. As a consequence, the agreements and alliances forged by the hegemonic power come unstuck. Global political and economic instability are thus likely to increase as the hegemon declines.

Robert Keohane expressly analyzes the consequences of hegemonic decline. He is most interested in assessing the explanatory power of neorealist arguments on hegemony (i.e., Krasner and Gilpin), though he also considers a liberal strand (Kindleberger) of the debate. More specifically, Keohane examines the impact of hegemonic decline on international regimes. All of the arguments he explicates, and Wallerstein as well, agree that as a hegemon loses its predominance, international agreements and conventions will lose their force and, possibly, collapse. Conversely, a hegemonic power is required to insure the international stability afforded by effective regimes. In a detailed study of three distinct regimes—international monetary relations, trade in manufactured goods, and the production and sale of petroleum—Keohane concludes that theories of hegemonic stability are only partially correct.

Keohane concedes that regimes are most orderly at moments of hegemony. But he finds differential rates and extents of change in regimes during the decade of U.S. hegemonic decline. Change is greatest in oil and least in manufacturing trade; transformation of the international monetary regime stands, qualitatively, somewhere between the other two. Keohane argues that, as currently constituted, hegemonic stability theory is unable to account for these differences. In seeking an alternative, he analyzes specific mechanisms of change for each regime. Only in the case of oil is an unambiguous link established between declining U.S. power and regime change. The Bretton Woods monetary system was vexed by the inherent instability of a gold-exchange system and particular failures of U.S. policy; declining hegemonic power was not the primary cause of regime change. Challenges to the liberal trade regime were rooted in the internal politics of advanced industrial countries. In sum, hegemonic decline does not have as widespread an impact as predicted.

Keohane suggests that an issue-specific revision of hegemonic stability theory would better explain the international regime changes he has found. Implicitly, such a revision turns hegemonic stability theory away from neorealism, which sees power distributed among discreet state actors, toward interdependence theory, which views power as diffused across different issue areas. Thus, Keohane supports a liberal interpretation of hegemony.

The Three Instances
of Hegemony in the History
of the Capitalist World-Economy

IMMANUEL WALLERSTEIN

When one is dealing with a complex, continuously evolving, large-scale historical system, concepts that are used as shorthand descriptions for structural patterns are useful only to the degree that one clearly lays out their purpose, circumscribes their applicability, and specifies the theoretical framework they presuppose and advance.

Let me therefore state some premises which I shall not argue at this point. If you are not willing to regard these premises as plausible, you will not find the way I elaborate and use the concept of hegemony very useful. I assume that there exists a concrete singular historical system which I shall call the "capitalist world-economy," whose temporal boundaries go from the long sixteenth century to the present. Its special boundaries originally included Europe (or most of it) plus Iberian America but they subsequently expanded to cover the entire globe. I assume this totality is a *system,* that is, that it has been relatively autonomous of external forces; or, to put it another way, that its patterns are explicable largely in terms of its internal dynamics. I assume that it is an *historical* system, that is, that it was born, has developed, and will one day cease to exist (through disintegration or fundamental transformation). I assume lastly that it is the dynamics of the system itself that explain its historically changing characteristics. Hence, insofar as it is a system, it has structures and these structures manifest themselves in cyclical rhythms, that is, mechanisms which reflect and ensure repetitious patterns. But insofar as this system is historical, no rhythmic movement ever returns the system to an

Reprinted from *International Journal of Comparative Sociology,* 24, 1–2 (1983), Immanuel Wallerstein, "Three Instances of Hegemony in the History of the Capitalist World Economy" by permission of E.J. Brill (USA), Inc.

equilibrium point but instead moves the system along various continua which may be called the secular trends of this system. These trends eventually must culminate in the impossibility of containing further reparations of the structured dislocations by restorative mechanisms. Hence the system undergoes what some call "bifurcating turbulence" and others the "transformation of quantity into quality."

To these methodological or metaphysical premises, I must add a few substantive ones about the operations of the capitalist world-economy. Its mode of production is capitalist; that is, it is predicated on the endless accumulation of capital. Its structure is that of an axial social division of labor exhibiting a core/periphery tension based on unequal exchange. The political superstructure of this system is that of a set of so-called sovereign states defined by and constrained by their membership in an interstate network or system. The operational guidelines of this interstate system include the so-called balance of power, a mechanism designed to ensure that no single state ever has the capacity to transform this interstate system into a single world-empire whose boundaries would match that of the axial division of labor. There have of course been repeated attempts throughout history of capitalist world-economy to transform it in the direction of a world-empire, but these attempts have all been frustrated. However, there have also been repeated and quite different attempts by given states to achieve hegemony in the interstate system, and these attempts have in fact succeeded on three occasions, if only for relatively brief periods.

The thrust of hegemony is quite different from the thrust to world-empire; indeed it is in many ways almost its opposite. I will therefore (1) spell out what I mean by hegemony, (2) describe the analogies in the three purported instances, (3) seek to decipher the roots of the thrust to hegemony and suggest why the thrust to hegemony has succeeded three times but never lasted too long, and (4) draw inferences about what we may expect in the proximate future. The point of doing all this is not to erect a Procrustean category into which to fit complex historical reality but to illuminate what I believe to be one of the central processes of the modern world-system.

1

Hegemony in the interstate system refers to that situation in which the ongoing rivalry between the so-called "great powers" is so unbalanced that one power is truly *primus inter pares;* that is, one power can largely impose its rules and its wishes (at the very least by effective veto power) in the economic, political, military, diplomatic, and even cultural arenas. The material base of such power lies in the ability of enterprises domiciled in that power to operate more efficiently in all three major economic areas—agro-industrial production, commerce, and finance. The edge in efficiency of which we are speaking is one so great that these enterprises can not only outbid enterprises domiciled in other great powers in the world market in general, but quite specifically in very many instances within the home markets of the rival powers themselves.

I mean this to be a relatively restrictive definition. It is not enough for one power's enterprises simply to have a larger share of the world market than any other or simply to have the most powerful military forces or the largest political role. I mean hegemony only to refer to situations in which the edge is so significant that allied major powers are *de facto* client states and opposed major powers feel relatively frustrated and highly defensive vis-à-vis the hegemonic power. And yet while I want to restrict my definition to instances where the margin or power differential is really great, I do not mean to suggest that there is ever any moment when a hegemonic power is omnipotent and capable of doing anything it wants. Omnipotence does not exist within the interstate system.

Hegemony therefore is not a state of being but rather one end of a fluid continuum which describes the rivalry relations of great powers to each other. At one end of this continuum is an almost even balance, a situation in which many powers exist, all somewhat equal in strength, and with no clear or continuous groupings. This is rare and unstable. In the great middle of this continuum, many powers exist, grouped more or less into two camps, but with several neutral or swing elements, and with neither side (nor *a fortiori* any single state) being able to impose its will on others. This is the statistically normal situation of rivalry within the interstate system. And at the other end lies the situation of hegemony, also rare and unstable.

At this point, you may see what it is I am describing but may wonder why I am bothering to give it a name and thereby focus attention upon it. It is because I suspect hegemony is not the result of a random reshuffling of the cards but is a phenomenon that emerges in specifiable circumstances and plays a significant role in the historical development of the capitalist world-economy.

2

Using this restrictive definition, the only three instances of hegemony would be the United Provinces in the mid-seventeenth century, the United Kingdom in the mid-nineteenth, and the United States in the mid-twentieth. If one insists on dates, I would tentatively suggest as maximal bounding points 1620–72, 1815–73, 1945–67. But of course, it would be a mistake to try to be too precise when our measuring instruments are both so complex and so crude.

I will suggest four areas in which it seems to me what happened in the three instances was analogous. To be sure, analogies are limited. And to be sure, since the capitalist world-economy is in my usage a single continuously evolving entity, it follows by definition that the overall structure was different at each of the three points in time. The differences were real, the outcome of the secular trends of the world-system. But the structural analogies were real as well, the reflection of the cyclical rhythms of this same system.

The first analogy has to do with the sequencing of achievement and loss of relative efficiencies in each of the three economic domains. What I believe occurred was that in each instance enterprises domiciled in the given power in

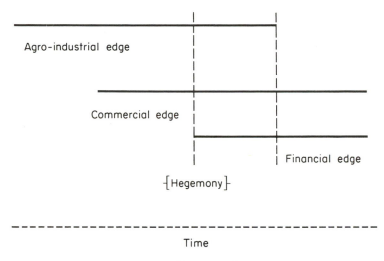

Figure 1. Economic position of hegemonic power

question achieved their edge first in agro-industrial production, then in commerce, and then in finance. I believe they lost their edge in this sequence as well (this process having begun but not yet having been completed in the third instance). Hegemony thus refers to that short interval in which there is *simultaneous* advantage in all three economic domains.

The second analogy has to do with the ideology and policy of the hegemonic power. Hegemonic powers during the period of their hegemony tended to be advocates of global "liberalism." They came forward as defenders of the principle of the free flow of the factors of production (goods, capital, and labor) throughout the world-economy. They were hostile in general to mercantilist restrictions on trade, including the existence of overseas colonies for the stronger countries. They extended this liberalism to a generalized endorsement of liberal parliamentary institutions (and a concurrent distaste for political change by violent means), political restraints on the arbitrariness of bureaucratic power, and civil liberties (and a concurrent open door to political exiles). They tended to provide a high standard of living for their national working classes, high by world standards of the time.

None of this should be exaggerated. Hegemonic powers regularly made exceptions to their anti-mercantilism, when it was in their interest to do so. Hegemonic powers regularly were willing to interfere with political processes in other states to ensure their own advantage. Hegemonic powers could be very repressive at home, if need be, to guarantee the national "consensus." The high working-class standard was steeply graded by internal ethnicity. Nevertheless, it is quite striking that liberalism as an ideology did flourish in these countries at precisely the moments of their hegemony, and to a significant extent only then and there.

The third analogy is in the pattern of global military power. Hegemonic powers were primarily sea (now sea/air) powers. In the long ascent to hegemony, they seemed very reluctant to develop their armies, discussing openly the potentially weakening drain on state revenues and manpower of becoming tied down in land wars. Yet each found finally that it had to develop a strong land army as well to face up to a major land-based rival which seemed to be trying to transform the world-economy into a world-empire.

In each case, the hegemony was secured by a thirty-year-long world war. By a world war, I shall mean (again somewhat restrictively) a land-based war that involves (not necessarily continuously) almost all the major military powers of the epoch in warfare that is very destructive of land and population. To each hegemony is attached one of these wars. World War Alpha was the Thirty Years' War from 1618 to 1648, where Dutch interests triumphed over Hapsburg in the world-economy. World War Beta was the Napoleonic Wars from 1792 to 1815, where British interests triumphed over French. World War Gamma was the long Euroasian wars from 1914 to 1945, where US interests triumphed over German.

While limited wars have been a constant of the operations of the interstate system of the capitalist world-economy (there having been scarcely any year when there was not some war somewhere within the system), world wars have been by contrast a rarity. In fact their rarity and the fact that the number and timing seems to have correlated with the achievement of hegemonic status by one power brings us to the fourth analogy.

If we look to those very long cycles that Rondo Cameron has dubbed "logistics," we can see that world wars and hegemony have been in fact related to them. There has been very little scholarly work done on these logistics. They have been most frequently discussed in the comparisons between the A–B sequences of 1100–1450 and 1450–1750. There are only a few discussions of the logistics that may exist after the latter point in time. But if we take the prime observation which has been used to define these logistics— secular inflation and deflation—the patterns seems in fact to have continued.

It therefore might be plausible to argue the existence of such (price) logistics up to today using the following dates: 1450–1730, with 1600–50 as a flat peak; 1730–1897, with 1810–17 as a peak; and 1897–?, with an as yet uncertain peak. If there are such logistics, it turns out that the world war and the (subsequent) hegemonic era are located somewhere around (just before and after) the peak of the logistic. That is to say, these processes seem to be the product of the long competitive expansion which seemed to have resulted in a particular concentration of economic and political power.

The outcome of each world war included a major restructuring of the interstate system (Westphalia; the Concert of Europe; the UN and Bretton Woods) in a form consonant with the need for relative stability of the now hegemonic power. Furthermore, once the hegemonic position was eroded economically (the loss of the efficiency edge in agro-industrial production), and therefore hegemonic decline set in, one consequence seemed to be the erosion of the alliance network which the hegemonic power had created patiently, and ultimately a serious reshuffling of alliances.

In the long period following the era of hegemony, two powers seemed eventually to emerge as the "contenders for the succession"—England and France after Dutch hegemony; the US and Germany after British; and now Japan and Western Europe after US. Furthermore, the eventual winner of the contending pair seemed to use as a conscious part of its strategy the gentle turning of the old hegemonic power into its "junior partner"—the English vis-à-vis the Dutch, the US vis-à-vis Great Britain . . . and now?

3

Thus far I have been primarily descriptive. I realize that this description is vulnerable to technical criticism. My coding of the data may not agree with everyone else's. I think nonetheless that as an initial effort this coding is defensible and that I have therefore outlined a broad repetitive pattern in the functioning of the interstate question. The question now is how to interpret it. What is there in the functioning of a capitalist world-economy that gives rise to such a cyclical pattern in the interstate system?

I believe this pattern of the rise, temporary ascendancy, and fall of hegemonic powers in the interstate system is merely one aspect of the central role of the political machinery in the functioning of capitalism as a mode of production.

There are two myths about capitalism put forward by its central ideologues (and strangely largely accepted by its nineteenth-century critics). One is that it is defined by the free flow of the factors of production. The second is that it is defined by the non-interference of the political machinery in the "market." In fact, capitalism is defined by the *partially* free flow of the factors of production and by the *selective* interference of the political machinery in the "market." Hegemony is an instance of the latter.

What defines capitalism most fundamentally is the drive for the endless accumulation of capital. The interferences that are "selected" are those which advance this process of accumulation. There are however two problems about "interference." It has a cost, and therefore the benefit of any interference is only a benefit to the extent it exceeds this cost. Where the benefits are available without any "interference," this is obviously desirable, as it minimizes the "deduction." And secondly, interference is always in favor of one set of accumulators as against another set, and the latter will always seek to counter the former. These two considerations circumscribe the politics of hegemony in the interstate system.

The costs to a given entrepreneur of state "interference" are felt in two main ways. First, in financial terms, the state may levy direct taxes which affect the rate of profit by requiring the firm to make payments to the state, or indirect taxes, which may alter the rate of profit by affecting the competitivity of a product. Secondly, the state may enact rules which govern flows of capital, labor, or goods, or may set minimum and/or maximum prices. While direct taxes always represent a cost to the entrepreneur, calculations concerning indirect taxes and state regulations are more complex, since they represent costs both to the entrepreneur and to (some of) his competitors. The chief concern in terms of individual accumulation is not the absolute cost of these measures but the comparative cost. Costs, even if high, may be positively desirable from the standpoint of a given entrepreneur, if the state's actions involve still higher costs to some competitor. Absolute costs are of concern only if the loss to the entrepreneur is greater than the medium-run gain which is possible through greater competitivity brought about by such state actions. It follows that absolute cost is of greatest concern to those entrepreneurs who would do best in open-market competition in the absence of state interference.

In general, therefore, entrepreneurs are regularly seeking state interference in the market in multiple forms—subsidies, restraints of trade, tariffs (which are penalties for competitors of different nationality), guarantees, maxima for input prices and minima for output prices, etc. The intimidating effect of internal and external repression is also of direct economic benefit to entrepreneurs. To the extent that the ongoing process of competition and state interference leads to oligopolistic conditions within state boundaries, more and more attention is naturally paid to securing the same kind of oligopolistic conditions in the most important market, the world market.

The combination of the competitive thrust and constant state interference results in a continuing pressure towards the concentration of capital. The benefits of state interference inside and outside the state boundaries is cumulative. In political terms, this is reflected as expanding world power. The edge a rising power's economic enterprises have vis-à-vis those of a competitive rising power may be thin and therefore insecure. This is where the world wars come in. The thirty-year struggle may be very dramatic militarily and politically. But the profoundest effect may be economic. The winner's economic edge is expanded by the very process of the war itself, and the postwar interstate settlement is designed to encrust that greater edge and protect it against erosion.

A given state thus assumes its world "responsibilities" which are reflected in its diplomatic, military, political, ideological, and cultural stances. Everything conspires to reinforce the cooperative relationship of the entrepreneurial strata, the bureaucratic strata, and with some lag the working-class strata of the hegemonic power. This power may then be exercised in a "liberal" form—given the real diminution of political conflict within the state itself compared to earlier and later periods, and to the importance in the interstate arena of delegitimizing the efforts of other state machineries to act against the economic superiorities of the hegemonic power.

The problem is that global liberalism, which is rational and cost-effective, breeds its own demise. It makes it more difficult to retard the spread of technological expertise. Hence over time it is virtually inevitable that entrepreneurs coming along later will be able to enter the most profitable markets with the most advanced technologies and younger "plant," thus eating into the material base of the productivity edge of the hegemonic power.

Secondly, the internal political price of liberalism, needed to maintain uninterrupted production at a time of maximal global accumulation, is the creeping rise of real income of both the working strata and the cadres located in the hegemonic power. Over time, this must reduce the competitivity of the enterprises located in this state.

Once the clear productivity edge is lost, the structure cracks. As long as there is a hegemonic power, it can coordinate more or less the political responses of all states with core-like economic activities to all peripheral states, maximizing thereby the differentials of unequal exchange. But when hegemony is eroded, and especially when the world-economy is in a Kondratieff downturn, a scramble arises among the leading powers for the smaller pie, which undermines their collective ability to extract surplus via unequal exchange. The rate of unequal exchange thereby diminishes (but never to zero) and creates further incentive to a reshuffling of alliance systems.

In the period leading to the peak of a logistic, which leads towards the creation of the momentary era of hegemony, the governing parable is that of the tortoise and the hare. It is not the state that leaps ahead politically and especially militarily that wins the race, but the one that plods along improving inch by inch its long-term competitivity. This requires a firm but discrete and intelligent organization of the entrepreneurial effort by the state-machinery. Wars may be left to others, until the climactic world war when the hegemonic power must at last invest its resources to clinch its victory. Thereupon comes "world responsibility" with its benefits but also its (growing) costs. Thus the hegemony is sweet but brief.

4

The inferences for today are obvious. We are in the immediate post-hegemonic phase of this third logistic of the capitalist world-economy. The US has lost its productive edge but not yet its commercial and financial superiorities; its military and political power edge is no longer so overwhelming. Its abilities to dictate to its allies (Western Europe and Japan), intimidate its foes, and overwhelm the weak (compare the Dominican Republic in 1965 with El Salvador today) are vastly impaired. We are in the beginnings of a major reshuffling of alliances. Yet, of course, we are only at the beginning of all this. Great Britain began to decline in 1873, but it was only in 1982 that it could be openly challenged by Argentina.

The major question is whether this third logistic will act itself out along the lines of the previous ones. The great difference is the degree to which the fact that the capitalist world-economy has entered into a structural crisis as an historical system will obliterate these cyclical processes. I do not believe it will obliterate them but rather that it will work itself out in part through them.

We should not invest more in the concept of hegemony than is there. It is a way of organizing our perception of process, not an "essence" whose traits are to be described and whose eternal recurrences are to be demonstrated and then anticipated. A processual concept alerts us to the forces at play in the system and the likely nodes of conflict. It does not do more. But it also does not do less. The capitalist world-economy is not comprehensible unless we analyze clearly what are the political forms which it has engendered and how these forms relate to other realities. The interstate system is not some exogenous, God-given variable which mysteriously restrains and interacts with the capitalist drive for the endless accumulation of capital. It is its expression at the level of the political arena.

The Theory of Hegemonic Stability and Changes in International Economic Regimes, 1967–1977

ROBERT O. KEOHANE

Background

In 1967 the world capitalist system, led by the United States, appeared to be working smoothly. Europe and Japan had recovered impressively from World War II and during the 1960s the United States had been enjoying strong, sustained economic growth as well. Both unemployment and inflation in seven major industrialized countries stood at an average of only 2.8 percent. International trade had been growing even faster than output, which was expanding at about 5 percent annually; and direct investment abroad was increasing at an even faster rate. The Kennedy round of trade talks was successfully completed in June 1967; in the same month, the threat of an oil embargo by Arab countries in the wake of an Israeli–Arab war had been laughed off by the Western industrialized states. Fixed exchange rates prevailed; gold could still be obtained from the United States in exchange for dollars; and a prospective "international money," Special Drawing Rights (SDRs) was created in 1967 under the auspices of the International Monetary fund (IMF). The United States, "astride the world like a colossus," felt confident enough of its power and position to deploy half a million men to settle the affairs of Vietnam. U.S. power and dynamism constituted the problem or the promise; "the American challenge" was global. Conservative and radical commentators alike regarded U.S. dominance as the central reality of contemporary world politics, although they differed as to whether its implications were benign or malign.

A decade later the situation was very different. Unemployment rates in the West had almost doubled while inflation rates had increased almost three-fold. Surplus capacity had appeared in the steel, textiles, and ship-building industries, and was feared in others. Confidence that Keynesian policies could ensure uninterrupted growth had been undermined if not shattered. Meanwhile, the United States had been defeated in Vietnam and no longer seemed to have either the capability or inclination to extend its military domination to the far corners of the world. The inability of the United States to prevent or counteract the oil price increases of 1973–74 seemed to symbolize the drastic changes that had taken place.

The decade after 1967 therefore provides an appropriate historical context for exploring recent developments in the world's political economy, and for testing some explanations of change.

What changes are observed, and how can they be accounted for in a politically sophisticated way? As this question suggests, this chapter has both a descriptive and an explanatory aspect. Descriptively, it examines changes between 1967 and 1977 in three issue areas: trade in manufactured goods, international monetary relations, and petroleum trade. The focus in each issue area is on the character of its *international regime*—that is, with the norms, rules, and procedures that guide the behavior of states and other important actors. In each issue area an international regime can be identified as of 1967; and in each area identifiable changes in that regime took place during the following decade. These changes in regimes constitute the dependent variable of this study.

The explanatory portion of this chapter attempts to test a theory of "hegemonic stability," which posits that changes in the relative power resources available to major states will explain changes in international regimes. Specifically, it holds that hegemonic structures of power, dominated by a single country, are most conducive to the development of strong international regimes whose rules are relatively precise and well obeyed. According to the theory, the decline of hegemonic structures of power can be expected to presage a decline in the strength of corresponding international economic regimes.

The Concept of International Regime

The concept of international regime can be relatively narrow and precise or quite elastic. Regimes in the narrow sense are defined by explicit rules, usually agreed to by governments at international conferences and often associated with formal international organizations. . . . The international monetary regime agreed to at Bretton Woods, which came fully into force at the end of 1958, was characterized by explicit rules mandating pegged exchange rates and procedures for consultation if exchange rates were to be changed. . . .

The nondiscriminatory reciprocal trade regime of the General Agreement on Tariffs and Trade (GATT) contains rules about which governmental measures affecting trade are permitted, and which are prohibited, by international agreement. . . .

The definition of regime employed in this chapter, however, is more elastic than this relatively precise and rule-oriented version. The focus is less on institutionalization and rule development than on patterns of regularized co-operative behavior in world politics. Therefore this chapter includes as regimes those arrangements for issue areas that embody implicit rules and norms insofar as they actually guide behavior of important actors in a particular issue area. The distinction between explicit and implicit rules is less important than the distinction between strong regimes—in which predictable, orderly behavior takes place according to a set of standards understood by participants—and weak ones—in which rules are interpreted differently or broken by participants. Explicit regimes may be stronger than implicit ones, but this is not always the case, as indicated by the weakness of the international monetary regime between 1971 and 1976 despite the fact that the rules of Bretton Woods still remained nominally in force.

The concept of international regime enables a coherent analysis of changes in world politics. Rather than on an explanation of particular events, in which idiosyncratic and frequently random factors have played a role, the focus is on a pattern of events—not on particular bargaining outcomes but on what a pattern of bargaining outcomes reveals about implicit norms and rules in world politics. . . . Having identified the international regime and described how it has changed, we can then proceed to the second, more difficult analytical task—to account for these changes in terms of deeper political and economic forces. Describing changes in regimes provides interpretive richness for the analysis of political behavior; attempts to explain these changes many lead to insights about causal patterns.

Explaining Changes in International Regimes

. . . . A parsimonious theory of international regime change has recently been developed by a number of authors, notably Charles Kindelberger, Robert Gilpin, and Stephen Krasner. According to this theory, strong international economic regimes depend on hegemonic power. Fragmentation of power between competing countries leads to fragmentation of the international economic regime; concentration of power contributes to stability. Hegemonic powers have the capabilities to maintain international regimes that they favor. They may use coercion to enforce adherence to rules; or they may rely largely on positive sanctions—the provision of benefits to those who cooperate. Both hegemonic powers and the smaller states may have incentives to collaborate in maintaining a regime—the hegemonic power gains the ability to shape and dominate its international environment, while providing a sufficient flow of benefits to small and middle powers to persuade them to acquiesce. Some international regimes can be seen partially as collective goods whose benefits (such as stable money) can be consumed by all participants without detracting from others' enjoyment of them. Insofar as this is the case, economic theory leads us to expect that extremely large, dominant countries will be particularly willing to provide these goods, while relatively small participants will attempt to secure "free rides" by avoiding proportionate shares of payment. International systems with highly skewed distributions of capabilities will therefore tend to be more amply supplied with such collective goods than systems characterized by equality among actors.

The particular concern of this chapter is the erosion of international economic regimes. The hegemonic stability theory seeks sources of erosion in changes in the relative capabilities of states. As the distribution of tangible resources, especially economic resources, becomes more equal, international regimes should weaken. One reason for this is that the capabilities of the hegemonial power will decline—it will become less capable of enforcing rules against unwilling participants, and it will have fewer resources with which to entice or bribe other states into remaining within the confines of the regime. Yet the incentives facing governments will also change. As the hegemonial state's margin of resource superiority over its partners declines, the costs of leadership will become more burdensome. Enforcement of rules will be more difficult and side payments will seem less justifiable. Should other states—now increasingly strong economic rivals—not have to contribute their "fair shares" to the collective enterprise? The hegemon (or former hegemon) is likely to seek to place additional burdens on its allies. At the same time, the incentives of the formerly subordinate secondary states will change. They will not only become more capable of reducing their support for the regime; they may acquire new interests in doing so. On the one hand, they will perceive the possibility of rising above their subordinate status, and they may even glimpse the prospect of reshaping the international regime in order better to suit their own interests. On the other hand, they may begin to worry that their efforts (and those of others) in chipping away at the hegemonial power and its regime may be too successful—that the regime itself may collapse. This fear, however, may lead them to take further action to hedge their bets, reducing their reliance on the hegemonical regime and perhaps attempting to set up alternative arrangements of their own.

As applied to the last century and a half, this theory—which will be referred to as the "hegemonic stability" theory—does well at identifying apparently necessary conditions for strong international economic regimes, but poorly at establishing sufficient conditions. International economic regimes have been most orderly and predictable where there was a single hegemonic state in the world system: Britain during the mid-nineteenth century in trade and until 1914 in international financial affairs; the United States after 1945. Yet although tangible U.S. power resources were large during the interwar period, international economic regimes were anything but orderly. High inequality of capabilities was not, therefore, a sufficient condition for strong international regimes; there was in the case of the United States a lag between its attainment of capabilities and its acquisition of a willingness to exert leadership, or of a taste for domination depending on point of view.

The concern here is not with the validity of the hegemonic stability theory throughout the last 150 years, but with its ability to account for changes in international economic regimes during the decade between 1967 and 1977. Since the United States remained active during those years as the leading capitalist country, the problems of "leadership lag" does not exist, which raises difficulties for the interpretation of the inter-war period. Thus the theory should apply to the 1967–1977 period. Insofar as "potential economic power" (Krasner's term) became more equally distributed—reducing the share of the United States—during the 1960s and early to mid-1970s, U.S.-created and U.S.-centered international economic regimes should also have suffered erosion or decline.

The hegemonic stability thesis is a power-as-resources theory, which attempts to link tangible state capabilities (conceptualized as "power resources") to behavior. In its simplest form, it is a "basic force model" in which outcomes reflect the potential power (tangible and known capabilities) of actors. Basis force models typically fail to predict accurately particular political outcomes, in part because differential opportunity costs often lead competing actors to use different proportions of their potential power. Yet they offer clearer and more easily interpretable explanations than "force activation models," which incorporate assumptions about differential exercise of power. Regarding tendencies rather than particular decisions, they are especially useful in establishing a baseline, a measure of what can be accounted for by the very parsimonious theory that tangible resources are directly related to outcomes, in this case to the nature of international regimes. The hegemonic stability theory, which is systemic and parsimonious, therefore seems to constitute a useful starting point for analysis, on the assumption that it is valuable to see how much can be learned from simple explanations before proceeding to more complex theoretical formulations.

Changes in International Economic Regimes, 1967–1977

The dependent variable in this analysis is international regime change between 1967 and 1977 in three issue areas: international monetary relations, trade in manufactured goods, and the production and sale of petroleum. . . . Descriptive contentions about international regime change in the three chosen areas are that (1) all three international regimes existing in 1967 became weaker during the subsequent decade; (2) this weakening was most pronounced in the petroleum area and in monetary relations, where the old norms were destroyed and very different practices emerged—it was less sudden and less decisive in the field of trade; and (3) in the areas of trade and money, the dominant political coalitions supporting the regime remained largely the same, although in money certain countries (especially Saudi Arabia) were added to the inner "club," whereas in the petroleum issue area, power shifted decisively from multinational oil companies and governments of major industrialized countries to producing governments. Taking all three dimensions into account, it is clear that regime change was most pronounced during the decade in oil and least pronounced in trade in manufactured goods, with the international monetary regime occupying an intermediate position.

The international trade regime of the General Agreement on Tariffs and Trade was premised on the principles of reciprocity, liberalization, and nondiscrimination. Partly as a result of its success, world trade had increased since 1950 at a much more rapid rate than world production. Furthermore, tariff liberalization was continuing: in mid-1967 the Kennedy round was successfully completed, substantially reducing tariffs on a wide range of industrial products. Yet despite its obvious successes, the GATT trade regime in 1967 was already showing signs of stress. The reciprocity and nondiscrimination provisions of GATT were already breaking down. Tolerance for illegal trade restrictions had grown, few formal complaints were being processed, and by 1967, GATT did not even require states maintaining illegal quantitative restrictions to obtain formal waivers of the rules. The "general breakdown in GATT legal affairs" had gone very far indeed, largely as a result of toleration of illegal restrictions such as the variable levy of the European Economic Community (EEC), EEC association agreements, and export subsidies. In addition, nontariff barriers, which were not dealt with effectively by GATT codes, were becoming more important. The trade regime in 1967 was thus strongest in the area of tariff liberalization, but less effective on nontariff barriers or in dealing with discrimination.

In the decade ending in 1967 the international monetary regime was explicit, formally institutionalized, and highly stable. Governments belonging to the International Monetary Fund were to maintain official par values for their currencies, which could be changed only to correct a "fundamental disequilibrium" and only in consultation with the IMF. During these nine years, the rules were largely followed, parity changes for major currencies were few and minor. . . . Nevertheless, as in trade, signs of weakness in the system were apparent. U.S. deficits had to a limited extent already undermined confidence in the dollar; and the United States was fighting a costly war in Vietnam which it was attempting to finance without tax increases at home. Consequentially, inflation was increasing in the United States.

The international regime for oil was not explicitly defined by intergovernmental agreement in 1967. There was no global international organization supervising the energy regime. Yet, as mentioned above, the governing arrangements for international oil production and trade were rather clear. With the support of their home governments, the major international oil companies cooperated to control production and, within limits, price. The companies were unpopular in the host countries, and these host country governments put the companies on the defensive on particular issues, seeking increased revenue or increased control. . . . The companies retained superior financial resources and capabilities in production, transportation, and marketing that the countries could not attain. Furthermore, the companies possessed superior information. . . .

Although the U.S. government did not participate directly in oil production or trade, it was the most influential actor in the system. The United States had moved decisively during and after World War II to ensure that U.S. companies would continue to control Saudi Arabian oil. Later, when the Anglo–Iranian Oil Company became unwelcome as sole concessionaire in Iran, the United States sponsored an arrangement by which U.S. firms received 40 percent of the consortium established in the wake of the U.S.-sponsored coup that overthrew Premier Mossadegh and restored the shah to his throne. U.S. tax policy was changed in 1950 to permit U.S. oil companies to increase payments to producing governments without sacrificing profits, thus solidifying the U.S. position in the Middle East and Venezuela. . . . The governing arrangements for oil thus reflected the U.S. government's interests in an ample supply of oil at stable or declining prices, close political ties with conservative Middle Eastern governments, and profits for U.S.-based multinational companies.

The international economic regime of 1977 looked very different. Least affected was the trade regime, although even here important changes had taken place. Between 1967 and 1977, nontariff barriers to trade continued to proliferate, and the principle of nondiscrimination was further undermined. Restrictions on textile imports from less developed countries, originally limited to cotton textiles, were extended to woolen and manmade fabrics in 1974. Nontariff barriers affecting world steel trade in the early 1970s included import licensing, foreign exchange restrictions, quotas, export limitations, domestic-biased procurement, subsidies, import surcharges, and antidumping measures. During the 1970s, "voluntary" export restraints, which had covered about one-eighth of U.S. imports in 1971 were further extended. . . . Contemporaneously, the European Economic Community launched an ambitious program to protect and rationalize some of its basic industries afflicted with surplus but relatively inefficient capacity, such as steel and shipbuilding. . . . The stresses on the international trade system, according to the director-general of GATT, "have now become such that they seriously threaten the whole fabric of postwar cooperation in international trade policy."

Nevertheless, the weakening of some aspects of the international trade regime had not led, by the end of 1977, to reductions in trade or to trade wars; in fact, after a 4 percent decline in 1975, the volume of world trade rose 11 percent in 1976 and 4 percent in 1977. Furthermore, by 1977 the Tokyo round of trade negotiations was well underway; in 1979 agreement was reached on trade liberalizing measures that not only would (if put into effect) reduce tariffs on industrial products, but that would also limit or prohibit a wide range on nontariff barriers, including export subsidies, national preferences on government procurement, and excessively complex import licensing procedures. The weakening of elements of the old regime was therefore accompanied both by expanding trade (although at a lower rate than before 1973) and by efforts to strengthen the rules in a variety of areas.

By 1977 the international monetary regime had changed much more dramatically. The pegged-rate regime devised at Bretton Woods had collapsed in 1971, and its jerry-built successor had failed in 1973. Since then, major currencies had been floating against one another, their values affected both by market forces and frequently extensive governmental intervention. . . . Exchange rates have fluctuated quite sharply at times, and have certainly been more unpredictable than they were in the 1960s. . . .

In the oil area, the rules of the old regime were shattered between 1967 and 1977, as power shifted dramatically from the multinational oil companies and home governments (especially the United States and Britain), on the one hand, to producing countries' governments, on the other. The latter, organized since 1960 in the Organization of Petroleum Exporting Countries, secured a substantial price rise in negotiations at Teheran in 1971, then virtually quadrupled prices without negotiation after the Yom Kippur War of October 1973. Despite some blustering and various vague threats the United States could do little directly about this, although high rates of inflation in industrial countries and the decline of the dollar in 1977 helped to reduce substantially the real price of oil between 1974 and 1977. By 1977 the United States had apparently conceded control of the regime for oil pricing and production to OPEC, and particularly to its key member, Saudi Arabia. OPEC made the rules in 1977, influenced (but not controlled) by the United States. Only in the case of a crippling supply embargo would the United States be likely to act. The United States was still, with its military and economic strength, an influential actor, but it was no longer dominant.

Reviewing this evidence about three international economic regimes supports the generalizations offered earlier. Although all three old regimes became weaker during the decade, this was most pronounced for oil and money, least for trade. In oil, furthermore, dominant coalitions changed as well, so that by 1977 the regime that existed, dominated by OPEC countries, was essentially a *new regime*. The old petroleum regime had disappeared. By contrast, the 1977 trade regime was still a recognizable version of the regime existing in 1967; and the international monetary regime of 1977, although vastly different than in 1967, retained the same core of supportive states along with the same international organization, the IMF, as its monitoring agent. Since the rules had changed, the function of the IMF had also changed; but it persisted as an element, as well as a symbol, of continuity. In the oil area, the emergence of the International Energy Agency (IEA) after the oil embargo symbolized discontinuity: only after losing control of the pricing-production regime did it become necessary for the industrialized countries to construct their own formal international organization.

The Theory of Hegemonic Stability and International Regime Change

It should be apparent from the above account that a theory purporting to explain international economic regime change between 1967 and 1977 faces two tasks: first, to account for the *general pattern* of increasing weakness, and second, to explain why the oil regime experienced the most serious changes, followed by money and trade. Furthermore, the hegemonic stability theory must show not only a correspondence between patterns of regime change and changes in tangible power resources, but it must be possible to provide at least a plausible account of how those resource changes could have caused the regime changes that we observe.

The most parsimonious version of a hegemonic stability theory would be that changes in the overall international economic structure account for the changes in international regimes that we have described. Under this interpretation, a decline in U.S. economic power (as measured crudely by gross domestic product) would be held reesponsible for changes in international economic regimes. Power in this view would be seen as a fungible set of tangible economic resources that can be used for a variety of purposes in world politics.

There are conceptual as well as empirical problems with this parsimonious overall structure theory. The notion that power resources are fungible—that they can be allocated to issues as policymakers choose, without losing efficacy—is not very plausible in world political economy. As David Baldwin has recently argued, this theory fails to specify the context within which specific resources may be useful: "What functions as a power resource in one policy-contingency framework may be irrelevant in another. The only way to determine whether something is a power resource or not is to place it in the context of a real or hypothetical policy-contingency framework."[1] A second problem with the overall structure version of the hegemonic stability thesis is itself contextual: since we have to account not only for the general pattern of increasing weakness but also for differential patterns by issue area, focusing on a single dependent variable will clearly not suffice. Changes in the overall U.S. economic position will clearly not explain different patterns of regime change in different issue areas.

To explain different patterns of regime change in different issue areas, a differentiated, issue-specific version of the hegemonic stability thesis has greater value than the overall structure version. According to this view, declines in resources available to the United States for use in a given issue area should be closely related to the weakening of the international regime (*circa* 1967) in that area. Specifically, the least evidence of structural change should be found in the trade area, an intermediate amount in international monetary relations, and the most in petroleum. This correspondence between changes in the independent and dependent variables, would lend support to the theory. To establish the theory of a firmer basis, however, it would be necessary to develop a plausible causal argument based on the hegemonic stability theory for the issue areas and regimes under scrutiny here.

The figures on economic resources provide prima facie support for the hegemonic stability thesis. The U.S. proportion of trade, for the top five market-economy countries, fell only slightly between 1960 and 1975. . . . As we saw, the international trade regime—already under pressure in 1967— changed less in the subsequent decade than the regimes for money and oil.

Table 1 (1–A). Distribution of Economic Resources, by Issue Area,
Among the Five Major Market-Economy Countries, 1960–1975

A. Trade Resources (Exports Plus Imports as Percentage of World Trade)

Year	United States	Germany	Britain	France	Japan	U.S. as Percent of Top Five Countries
1960	13.4	8.1	8.7	4.9	3.3	35
1965	14.4	9.4	8.0	5.7	4.2	35
1970	15.0	11.0	6.9	6.3	6.2	33
1975	13.0	10.0	5.8	6.4	6.6	31

Source: Kenneth N. Waltz, *Theory of International Politics* (Reading, Mass.: Addison-Wesley, 1979), Appendix Table IV, p. 215.

Last column calculated from these figures.

Table 1 (1–B). Distribution of Economic Resources, by Issue Area,
Among the Five Major Market-Economy Countries, 1960–1975

Monetary Resources (Reserves as Percentage of World Reserves)

Year	United States	Germany	Britain	France	Japan	U.S. as Percent of Top Five Countries
1960	32.4	11.8	6.2	3.8	3.3	56
1965	21.8	10.5	4.2	9.0	3.0	45
1970	15.5	10.7	3.0	5.3	5.2	39
1975	7.0	13.6	2.4	5.5	5.6	21

Source: Calculated from International Financial Statistics (Washington: IMF), Volume XXXI–5 (May, 1978), 1978 Supplement, pp. 34–35.

Last column calculated from these percentages.

U.S. financial resources in the form of reserves fell sharply, reflecting the shift from U.S. dominance in 1960 to the struggles over exchange rates of the 1970s. In view of the continued ability of the United States to finance its deficits with newly printed dollars and treasury bills rather than with reserves, Table (1–C) should not be overinterpreted: it does *not* mean that Germany was "twice as powerful as the United States" in the monetary area by 1975. Yet it does, as indicated above, signal a very strong shift in the resource situation of the United States. Finally, the petroleum figures—especially in Table 1–C—are dramatic: the United States went from a large positive position in 1956 and a small positive position in 1967 to a very large petroleum deficit by 1973. The hegemonic stability theory accurately predicts from these data that U.S. power in the oil area and the stability of the old international oil regime would decline sharply during the 1970s.

These findings lend plausibility to the hegemonic stability theory by not disconfirming its predictions. They do not, however, establish its validity, even for this limited set of issues over one decade. It is also necessary, before

Table 1 (1–C.1). Distribution of Economic Resources, by Issue Area, Among the Five Major Market-Economy Countries, 1960–1975

C. Petroleum Resources
1. United States Imports and Excess Production Capacity in Three Crisis Years

Year	U.S. Oil Imports as Percent of Oil Consumption	U.S. Excess Production Capacity as Percent Oil Consumption	Net U.S. Position
1956	11	25	+14
1967	19	25	+6
1973	35	10	−25

Source: Joel Darmstadter and Hans H. Landsberg, "The Economic Background," in Raymond Vernon, ed., *The Oil Crisis,* special issue of *Daedalus,* Fall, 1975 (pp. 30–31).

Table 1 (C.2). Distribution of Economic Resources, by Issue Area, Among the Five Major Market-Economy Countries, 1960–1975

2. Oil Imports as Percentage of Energy Supply

Year	United States	Western Europe	Japan	Ratio of U.S. to European Dependence
1967	9	50	62	.18
1970	10	57	73	.18
1973	17	60	80	.28
1976	20	54	74	.37

Source: Kenneth N. Waltz, *Theory of International Politics* (Reading, Mass.: Addison-Wesley, 1979) Appendix Table X, p. 221. Last column calculated from these figures.

concluding that the theory accounts for the observed changes, to see whether plausible causal sequences can be constructed linking shifts in the international distribution of power to changes in international regimes. The following sections of this paper therefore consider the most plausible and well-founded particular accounts of changes in our three issue areas, to see whether the causal arguments in these accounts are consistent with the hegemonic stability theory. The ensuing discussion begins with oil, since it fits the theory so well, and then addresses the more difficult cases.

Interpreting Changes in the Petroleum Regime

The transformation in oil politics between 1967 and 1977 resulted from a change in the hegemonic coalition making the rules and supporting the regime: OPEC countries, particularly Saudi Arabia, replaced the Western powers, led by the United States. OPEC members had previously lacked the ability to capture monopolistic profits by forming a producers' cartel. In part, this reflects low self-confidence. . . . Yet OPEC's impotence was also a result, in the 1950s and 1960s, of overwhelming U.S. power. Until the huge asymmetry between U.S. power and that of the OPEC members was reduced or reversed, massive changes in the implicit regime could not be expected to occur. Without these changes, neither foolish U.S. tactics nor an Israeli–Arab war could have led to the price rises observed in February 1971 or October– December 1973.

U.S. military power vis-à-vis Middle Eastern members of OPEC was lower in the late 1960s and early 1970s than it had been before the entry of Russia into the Middle East in 1955; but it was not clear that U.S. military power declined dramatically between 1967 and the oil crisis of 1973. Yet fundamental shifts in available petroleum supplies were taking place. When previous oil crises had threatened in the wake of Arab–Israeli wars in 1956 and 1967, the United States was able to compensate by increasing domestic production, since about 25 percent of its oil-producing capacity was not being used prior to each crisis. In 1973 U.S. spare capacity had declined to about 10 percent of the total. In 1956 U.S. imports were only about 11 percent of consumption, mostly from Venezuela and Canada; in 1967 they constituted 19 percent; but by 1973 they amounted to over 35 percent, a substantial proportion of which came from the Middle East. U.S. proved reserves had fallen from 18.2 to 6.4 percent of world proved reserves. In the earlier situations, the United States could be "part of the solution;" in 1973, it was "part of the problem." Its fundamental petroleum resource base had been greatly weakened.

The hegemonic stability model leads us to expect a change in international petroleum arrangements during the mid-1970s: the dominance of the United States and other industrialized countries was increasingly being undermined, as OPEC members gained potential power resources at their expense. What the Yom Kippur War did was to make the Arab members of OPEC willing to take greater risks. When their actions succeeded in quadrupling the price of oil almost overnight, mutual confidence rose that members of the cartel who cut back production would not be "double-crossed" by other producers, but would rather benefit from the externalities (high prices as a result of supply shortages) created by others' similar actions. Calculations about externalities became positive and risks fell. A self-reinforcing cycle of underlying resource strength leading to success, to increased incentives to cooperate, and to greater strength was launched.

Interpreting Changes in the International Monetary Regime

The breakdown of the Bretton Woods pegged-rate monetary system is usually attributed by economists principally to two factors. First is the inherent instability of a gold-exchange standard, which Benjamin J. Cohen describes, with reference to the 1960s, as follows:

> A gold-exchange standard is built on the illusion of convertibility of its fiduciary element into gold at a fixed price. The Bretton Woods system, though,

was relying on deficits in the U.S. balance of payments to avert a world liquidity shortage. Already, America's "overhang" of overseas liabilities to private and official foreigners was growing larger than its gold stock at home. The progressive deterioration of the U.S. net reserve position, therefore, was bound in time to undermine global confidence in the dollar's continued convertibility. In effect, governments were caught on the horns of a dilemma. To forestall speculation against the dollar, U.S. deficits would have to cease. But this would confront governments with the liquidity problem. To forestall the liquidity problem, U.S. deficits would have to continue. But this would confront governments with the confidence problem. Governments could not have their cake and eat it too.[2]

This situation would have made the international monetary system of the 1960s quite delicate under the best of circumstances. . . . Rather than eliminating this deficit, the United States let its balance of payments on current account deteriorate sharply in the last half of the 1960s, as a result of increased military spending to fight the Vietnam War, coupled with a large fiscal deficit, excess demand in the United States, and the inflationary momentum that resulted. When U.S. monetary policy turned from restriction to ease in 1970, in reaction to a recession, huge capital outflows took place. The U.S. decision of August 1971 to suspend the convertibility of the dollar into gold and thus to force a change in the Bretton Woods regime followed. An agreement reached in December 1971 to restore fixed exchange rates collapsed under the pressure of monetary expansion in the United States and abroad and large continuing U.S. deficits, which reached a total (an official settlements basis) of over $50 billion during the years 1970–73.

As this account suggests, the collapse of the international monetary regime in 1971–1973 was in part a result of the inherent instability of gold-exchange systems; but the proximate cause was U.S. economic policy devised in response to the exigencies of fighting an unpopular and costly war in Vietnam. The hegemonic stability theory does not account for either the long-run entropy of regimes resting on both gold and foreign exchange or policy failures by the U.S. government. Yet in at least a narrow sense the theory is consistent with events: the Bretton Woods regime collapsed only after U.S. reserves had fallen sharply, which contributed to the difficulty of maintaining the value of the dollar at the old exchange rates.

The hegemonic stability theory is thus not disconfirmed by the monetary case. For several reasons, however, it functions as a highly unsatisfactory explanation of regime change in the monetary area. In the first place, the resources that were most important to the United States to maintain the regime were not tangible resources (emphasized by the theory) but the symbolic resources that go under the name of "confidence" in discussion of international financial affairs. U.S. reserves were less important than confidence in U.S. policy: as a reserve-currency country, the United States could generate more international money (dollars) as long as holders of dollars believed that the dollar would retain value compared to alternative assets, such as other currencies or gold. By 1970–71, however, confidence in U.S. economic policy, and hence in the dollar, had become severely undermined, and after August 1971 became impossible to restore.

The second major problem with the hegemonic stability theory's explanation of the monetary case is that it focuses on a variable—U.S. resources in the monetary area—which was itself largely a result of U.S. policy. The theory, as indicated earlier is systemic: but the most important sources of change (*reflected* in resource shifts) lay within the U.S. polity. To some extent, of course, this was true in the oil areas as well, since the United States could have conserved its oil resources during the 1950s and 1960s; but it is particularly important in monetary politics, since confidence depended on evaluations of U.S. policy and expectations about it. Perceptions of U.S. economic policy as inflationary thus translated directly into loss of intangible U.S. resources (confidence on the part of foreigners) through changes in the expectations of holders, or potential holders of dollars.

The final problem with the explanation offered by the hegemonic stability theory is that it does not capture the *dual* nature of the U.S. power position in 1971. On the one hand, as we have seen, the U.S. position was eroding. Yet to a considerable extent the scope of U.S. weakness was the *result* of the rules of the old regime: within these rules the United States, not being able to force creditor nations such as Germany and Japan to revalue their currencies, thus was in the position of having to defend the dollar by a variety of short-term expedients. Only by breaking the rules explicitly—by suspending convertibility of the dollar into gold—could the United States transform the bargaining position and make its creditors offer concessions of their own. . . . The United States thus had a strong political incentive to smash the old regime, and it also had the political power to do so. Once the regime had been destroyed, other governments had to heed U.S. wishes, since the active participation of the United States was the sine qua non of a viable international monetary regime.

. . . . Had its own economic policies been tailored to international demands, the United States could probably in 1971 have resumed leadership of a reconstructed international monetary system; but the United States did not have sufficient power to compel others to accept a regime in which only it would have monetary autonomy. Between 1971 and 1976, the United States was the most influential actor in international monetary negotiations, and secured a weak flexible exchange rate regime that was closer to its own preferences than to those of its partners; but given its own penchant for monetary autonomy, it could not construct a strong, stable new regime.

Interpreting Changes in the International Trade Regime

As has been seen, changes in the trade regime between 1967 and 1977 were broadly consistent with changes in potential power resources in the issue area. Power resources (as measured by shares of world trade among the industrialized countries) changed less in trade than in money or oil; and the regime changed less as well. So once again, the hegemonic stability theory is not disconfirmed.

The causal argument of the hegemonic stability theory, however, implies that the changes we do observe in trade (which are less than those in money and oil but are by no means insignificant) should be ascribable to changes in international political structure. Yet this does not appear to be the case. Protectionism is largely a grass-roots phenomenon, reflecting the desire of individuals for economic security and stability and of privileged groups for higher incomes than they would command in a free market. . . . Most governments of advanced capitalist states show little enthusiasm for protectionist policies, but have been increasingly goaded into them by domestic interests.

A recent GATT study identifies as a key source of protectionism "structural weaknesses and maladjustments" in the countries of the Organization for Economic Cooperation and Development (OECD). Its authors focus particularly on the recent tendency in Europe for wages to rise at similar rates across sectors, regardless of labor productivity. Yet industries that are old in one country, faced by dynamic, low-cost competitors from abroad, can only adjust effectively either by paying lower wages than higher-productivity industries or by reducing employment: "To maintain *both* the wage differential and the absolute size of employment in the industry, protection is necessary." Yet the workers affected and their political leaders may prevent steps to widen wage differentials or to reduce employment. This resistance to change leads both to inflation and to pressures for protection. According to this view, governments share the responsibility for inflation and protectionism insofar as they tolerate and accommodate these pressures.[3]

It may be true that for reasons internal to advanced industrial societies—or to certain societies, since some seem to be affected more than others—people of these countries are more resistant to adjustment than they once were. This may not just be a problem of the masses. More emphasis could be placed on insufficient levels of research and development by governments and firms, thus contributing to failure to innovate; or one could attempt to account for stagnation by reference to managerial inadequacy on the part of leaders of established oligopolistic industries. In either case, however, explanations from domestic politics or political culture would appear necessary to account for changes in international regimes. . . .

Most explanations of increased protectionism also focus on the recession of the 1970s and the rise of manufactured exports from less developed countries. Between the end of 1973 and the end of 1977, rates of growth in industrial production lagged throughout the OECD area, and in Western Europe and Japan industrial production hardly grew at all. Unemployment rates in Europe in 1977 were three times as high as average unemployment rates between 1957 and 1973. At the same time, recessionary pressure was accentuated by the rapid growth of exports of manufactured goods from developing countries, which increased by about 150 percent in volume terms between 1970 and 1977, while manufacturing output in the developed market economy countries increased by only about 30 percent. . . . Some recent restrictive measures, particularly the progressive tightening of export restraints on textiles, reflect these pressures from dynamic developing country exporters.

To some extent, difficulties in maintaining liberal trade among the OECD countries do reflect erosion of U.S. hegemony, although this is more pronounced as compared with the 1950s than with the late 1960s. In the 1950s the United States was willing to open its markets to Japanese goods in order to integrate Japan into the world economic system, even when most European states refused to do so promptly or fully. This has been much less the case in recent years. Until the European Common Market came into existence, the United States dominated trade negotiations; but since the EEC has been active, it has successfully demanded numerous exceptions to GATT rules. Relative equality in trade-related power resources between the EEC and the United States seems to have been a necessary, if not sufficient, condition for this shift.

On the whole, the hegemonic stability theory does not explain recent changes in international trade regimes as well as it explains changes in money or oil. The theory is not disconfirmed by the trade evidence, and correctly anticipates less regime change in trade than in money or oil; but it is also not very helpful in interpreting the changes that we do observe. Most major forces affecting the trade regime have little to do with the decline of U.S. power. For an adequate explanation of changes in trade, domestic political and economic patterns, and the strategies of domestic political actors, would have to be taken into account.

Hegemonic Stability and Complex Interdependence: A Conclusion

A structural approach to international regime change, differentiated by issue area, takes us some distance toward a sophisticated understanding of recent changes in the international politics of oil, money, and trade. Eroding U.S. hegemony helps to account for political reversals in petroleum politics, to a lesser extent for the disintegration of the Bretton Woods international monetary regime, and to a still lesser extent for the continuing decay of the GATT-based trade regime.

. . . . There is a definite correspondence between the expectations of a hegemonic stability theory and the evidence presented here. Changes in tangible power resources by issue area and changes in regimes tend to go together. In terms of causal analysis, however, the results are more mixed. In the petroleum area a plausible and compelling argument links changes in potential economic power resources directly with outcomes. With some significant caveats and qualifications, this is also true in international monetary politics; but in trade, the observed changes do not seem causally related to shifts in international political structure.

On the basis of this evidence, we should be cautious about putting the hegemonic stability theory forward as a powerful explanation of events. It is clearly useful as a first step; to ignore its congruence with reality, and its considerable explanatory power, would be foolish. Nevertheless, it carries with it the conceptual difficulties and ambiguities characteristic of power analysis. Power is viewed in terms of resources; if the theory is to be operationalized, these resources have to be tangible. Gross domestic product, oil import dependence, international monetary reserves, and share of world trade are crude indicators of power in this sense. Less tangible resources such as confidence (in oneself or in a currency) or political position relative to other actors are not taken into account. Yet these sources of influence would seem to be conceptually as close to what is meant by "power resources" as are the more tangible and measurable factors listed above. Tangible resource models, therefore, are inherently crude and can hardly serve as more than first-cut approximations—very rough models that indicate the range of possible behavior or the probable path of change, rather than offering precise predictions.

The version of the hegemonic stability theory that best explains international economic regime change between 1967 and 1977 is an "issue structure" rather than "overall structure" model. That is, changes in power resources specific to particular issue areas are used to explain regime change. Issue structure models such as this one assume the separateness of issue areas in world politics; yet functional linkages exist between issue areas, and bargaining linkages are often drawn by policymakers between issue areas that are not functionally linked. The decline in the value of the dollar form 1971–1973, after the conclusion of the Teheran Agreements in February 1971, contributed strongly to dissatisfaction by OPEC states with oil prices, since it adversely affected their share of the rewards. . . .

The existence of interissue linkages limits the explanatory power of issue structure models. To solve this problem would require a strong and sophisticated theory of linkage, which would indicate under what conditions linkages between issue areas would be important and what their impact on outcomes would be. No issue-specific explanation of events can be completely satisfactory in a world of multiple issues linked in a variety of ways.

Despite all the limitation on power structure analysis, beginning with it has the great advantage of setting up some very general predictions based on a theory that requires only small amounts of information. A remarkable portion of the observed changes rely only on this parsimonious, indeed, almost simple-minded theory. Furthermore, its very inadequacies indicate where other explanations and other levels of analysis must be considered. Having examined the explanatory strengths and weaknesses of a hegemonic stability theory, we understand better what puzzles remain to be solved by investigating other systemic theories or by focusing on domestic politics and its relationship to foreign economic policy. Beginning with a simple, international-level theory clarifies the issues. It helps to bring some analytical rigor and order into the analysis of international economic regimes. Without employing a structural model as a starting point, it is difficult to progress beyond potentially rich but analytically unsatisfactory description, which allows recognition of complexity to become a veil hiding our ignorance of the forces producing change in world political economy. To limit ambitions to such description would be a premature confession of failure.

Notes

1. David A. Baldwin, "Power Analysis and world Politics: New Trends vs. Old Tendencies," *World Politics* (January 1979):165.

2. Benjamin J. Cohen, *Organizing the World's Money,* p. 99.

3. Richard Blackhurst, Nicolas Marian, and Jan Tumlir, *Trade Liberalization, Protectionism and Interdependence,* GATT Studies in International Trade, no. 5 (Geneva: GATT, November 1977), esp. pp. 44–52. For a discussion along similar lines, see the "McCracken Report," esp. chaps. 5, 8.

9

Regime Analysis

Regime analysis, like the rational choice approach, emerged from liberal theory. It, too, has been integrated into other theoretical systems, especially neorealism, and has developed into a distinct subfield of IPE. Unlike rational choice, which is basically a methodological technique, regime analysis centers on hypothesized empirical relationships among global actors: international regimes. Thus, theoretical synthesis centered on regime analysis does not involve the same epistemological problems as rational choice. Although regime analysis is a product of liberal theory, it is not necessarily linked to a specific set of conceptual and methodological assumptions. It can be appropriated by contending schools of thought without threatening core theoretical propositions.

The regime analysis of Donald J. Puchala and Raymond F. Hopkins is consistent with the liberal tradition. They take up the time-honored liberal issue of the role of principles in international affairs. Their definition of regimes—"a set of principles, norms, rules and procedures around which actors' expectations converge"—is conventional but their application of the concept is broader than other regime analysts. Puchala and Hopkins see normative forces at work in virtually every issue-area of international relations. This suggests that regime-based principles and norms have greater efficacy than realist theory would allow. Puchala and Hopkins argue that regime analysis, as such, is an improvement over cruder power politics variants of realism. They do not lapse into unwarranted idealism, but clearly resuscitate long-standing liberal positions.

The question for Puchala and Hopkins is, therefore, to explain why some regimes are more effective in constraining state behavior than others. They consider the extent to which certain internal characteristics influence regime effectiveness. Regimes may be specific to one issue or more diffuse, involving numerous related issues; they may be formal or informal; subject to evolutionary or revolutionary change; and distinguished by certain distributive biases. Puchala and Hopkins examine two regimes with very different configurations: colonialism (diffuse, informal, subject to revolutionary change, and very biased) and food (specific, formal, open to evolutionary change, and less biased) in an effort to clarify the significance of these variables.

Their conclusions are reminiscent of interdependence theory. Although internal characteristics are helpful in understanding the dynamics of specific regimes, on a more general level regime effectiveness is influenced by the broader international context. Regimes are likely to constrain state behavior when they mediate relations among countries of relatively equal power. Additionally, regimes are important in situations where power is diffused, as in international organizations, and at moments of power transitions. In short, the global distribution of power and the definition of national interests are not negated by the existence of regimes. Thus, Puchala and Hopkins, in a manner similar to Keohane and Nye's notion of complex interdependence, do not completely reject realist analysis; they qualify it. They argue that regimes "tend to have inertia or functional autonomy and continue to influence behavior even though their norms have ceased either to reflect the preferences of powers or to be buttressed by their capabilities." This implies that, under certain circumstances, state action may be significantly influenced by relatively autonomous international regimes, a point that undermines state-centric realist arguments.

Stephen Krasner's regime analysis is more sympathetic to realism. His argument both complements and counters Puchala and Hopkin's study. Krasner accepts the existence of regimes, though not as broadly as Hopkins and Puchala. However, he is concerned less with how regimes constrain state action than with how states, especially less developed countries (LDCs), use regimes to further their own interests. In this manner, he integrates the regime concept into neorealism.

Krasner argues that regimes are created by a hegemonic power to suit its particular interests. Paradoxically, to maintain international legitimacy the hegemon must grant regime institutions a certain operational autonomy from its domination. Krasner thus agrees with Hopkins and Puchala that regimes may diverge from great power control. Yet, instead of the implication that state interests are thereby limited by international regimes, Krasner suggests that regime autonomy provides an opening for other states, LDCs in particular, to redefine regimes in their own interests. Indeed, in light of LDC internal and international vulnerabilities, regimes provide weaker countries with a promising avenue of global influence. Regimes, from this perspective, are a form of "meta-power;" that is, they define how power will be exercised in certain issue areas. LDCs are weak in terms of "relational power," the ability to use tangible capabilities to influence outcomes, but, through concerted action, may affect the broader environment within which decisions are made, the realm of meta-power. Regime politics are, by this reckoning, a new form of the international "struggle for power."

Krasner's handling of regimes takes the state as the unit of analysis, dwelling at length on LDC national capabilities and motivations. He therefore preserves a key element of realist analysis. Alternatively, Puchala and Hopkins, in their consideration of internal regime characteristics, take the regime itself as the unit of analysis. Thus, Krasner uses regime analysis to explain LDC behavior from a neorealist point of view; Puchala and Hopkins employ regime analysis to discuss the transnationalization of power and interests. Krasner is also less optimistic regarding the potential for regime-based international cooperation. LDC attempts to exercise meta-power through regime change are driven by a rejection of existing principles and norms. Puchala and Hopkins suggest that regimes promote collective action by "constraining unilateral adventurousness or obduracy." Krasner, by contrast, pictures regimes as the battlefield of LDCs against global liberalism.

In sum, regime analysis has been employed by liberals and neorealists alike. Although obvious differences in interpretation exist, the regime concept is common ground between two distinct research programs; it provides a point of synthesis.

International Regimes:
Lessons from Inductive Analysis

DONALD J. PUCHALA AND RAYMOND F. HOPKINS

Rising interest in the concept "international regime" in the 1970s is much like that accorded to "international system" in the 1950s. It has become intellectually fashionable to speak and write about regimes. Current faddishness notwithstanding, the purpose of this article is to show that the notion of regime is analytically useful, and that the concept is therefore likely to become a lasting element in the theory of international relations. As realist and other paradigms prove too limited for explaining an increasingly complex, interdependent, and dangerous world, scholars are searching for new ways to organize intellectually and understand international activity. Using the term regime allows us to point to and comprehend sets of activities that might otherwise be organized or understood differently. Thinking in terms of regimes also alerts us to the subjective aspects of international behavior that might be overlooked altogether in more conventional inquiries.

A regime . . . is a set of principles, norms, rules, and procedures around which actors' expectations converge. These serve to channel political action within a system and give it meaning. For every political system, be it the United Nations, the United States, New York City, or the American Political Science Association, there is a corresponding regime. Regimes constrain and regularize the behavior of participants, affect which issues among protagonists move on and off agendas, determine which activities are legitimized or condemned, and influence whether, when, and how conflicts are resolved.

Several particular features of the phenomenon of regimes, as we conceive

Reprinted from *International Organization*, Vol. XXXVI, No. 2, Donald J. Puchala and Raymond F. Hopkins, "International Regimes: Lessons from Inductive Analysis," by permission of The MIT Press, Cambridge, MA.

of it, are worth noting, since other authors do not stress or, in the case of some, accept these points. We stress five major features.

First, a regime is an attitudinal phenomenon. Behavior follows from adherence to principles, norms, and rules, which legal codes sometimes reflect. But *regimes themselves are subjective:* they exist primarily as participants' understandings, expectations or convictions about legitimate, appropriate or moral behavior.

Second, an international regime includes tenets concerning appropriate procedures for making decisions. This feature, we suggest, compels us to identify a regime not only by a major substantive norm (as is done in characterizing exchange rate regimes as fixed or floating rate regimes) but also by the broad norms that establish procedures by which rules or policies—the detailed extensions of principles—are reached. Questions about the norms of a regime, then, include who participates, what interests dominate or are given priority, and what rules serve to protect and preserve the dominance in decision making.

Third, a description of a regime must include a characterization of the major principles it upholds (e.g., the sanctity of private property or the benefits of free markets) as well as the norms that prescribe orthodox and proscribe deviant behavior. It is especially useful to estimate the hierarchies among principles and the prospects for norm enforcement. These bear upon the potential for change.

Fourth, each regime has a set of elites who are the practical actors within it. Governments of nation-states are the prime official members of most international regimes, although international, transnational, and sometimes subnational organizations may practically and legitimately participate. More concretely, however, regime participants are most often bureaucratic units or individuals who operate as parts of the "government" of an international subsystem by creating, enforcing or otherwise acting in compliance with norms. Individuals and bureaucratic roles are linked in international networks of activities and communication. These individuals and rules govern issue-areas by creating and maintaining regimes.

Finally, a regime exists in every substantive issue-area in international relations where there is discernibly patterned behavior. Wherever there is regularity in behavior some kinds of principles, norms or rules must exist to account for it.

Regime Distinctions Important for Comparative Study

. . . . Our examination of several international regimes suggests four characteristics of theoretical importance.

1. Specific vs. Diffuse Regimes

. . . . Regimes can be differentiated according to function along a continuum ranging from specific, single-issue to diffuse, multi-issue. They may also be categorized by participants according to whether a few or a great many actors subscribe to their principles or at least adhere to their norms. No international regimes command universal adherence, though many approach it. More specific regimes often tend to be embedded in broader, more diffuse ones—the principles and norms of the more diffuse regimes are taken as given in the more specific regimes. In this sense we may speak of normative *superstructures,* which are reflected in functionally or geographically specific normative substructures or regimes. For example, in the nineteenth century, principles concerning the rectitude of the balance of power among major actors (the normative superstructure) were reflected in norms legitimizing and regulating colonial expansion (a substructure), and in those regulating major-power warfare (another substructure). . . .

2. Formal vs. Informal Regimes

Some regimes are legislated by international organizations, maintained by councils, congresses or other bodies, and monitored by international bureaucracies. We characterize these as "formal" regimes. The European Monetary System is one example. By contrast, other, more "informal" regimes are created and maintained by convergence or consensus in objectives among participants, enforced by mutual self-interest and "gentlemen's agreements," and monitored by mutual surveillance. For example, Soviet–American detente between 1970 and 1979 could be said to have been governed by a regime that constrained competitiveness and controlled conflict in the perceived mutual interests of the superpowers. Yet few rules of the relationship were ever formalized and few institutions other than the Hot Line and the Helsinki accords were created to monitor and enforce them.

3. Evolutionary vs. Revolutionary Change

Regimes change substantively in at least two different ways: one preserves norms while changing principles; the other overturns norms in order to change principles. Regimes may change qualitatively because those who participate in them change their minds about interests and aims, usually because of changes in information available to elites or new knowledge otherwise attained. We call this *evolutionary change,* because it occurs within the procedural norms of the regime, usually without major changes in the distribution of power among participants. Such change, undisturbing to the power structure and within the regime's "rules of the game," is rather exceptional and characteristic mainly of functionally specific regimes.

By contrast, *revolutionary* change is more common. Most regimes function to the advantage of some participants and the disadvantage of others. The disadvantaged accept regime principles and norms (and diminished rewards or outright penalties) because the costs of noncompliance are understood to be higher than the costs of compliance. But disadvantaged participants tend to formulate and propagate counterregime norms, which either circulate in the realm of rhetoric or lie dormant as long as those who dominate the existing regime preserve their power and their consequent ability to reward compliance and punish deviance. However, if and when the power structure alters, the normative contents of a prevailing regime fall into jeopardy. Power transition ushers in regime transformation; previously disadvantaged but newly powerful participants ascend to dominance and impose new norms favoring their own interests. In extreme cases the advantaged and disadvantaged reverse status, and a new cycle begins with regime change contingent upon power change. Such revolutionary change is more characteristic of diffuse regimes, highly politicized functional regimes, or those where distributive bias is high.

4. Distributive Bias

All regimes are biased. They establish hierarchies of values, emphasizing some and discounting others. They also distribute rewards to the advantage of some and the disadvantage of others, and in so doing they buttress, legitimize, and sometimes institutionalize international patterns of dominance, subordination, accumulation, and exploitation. In general, regimes favor the interests of the strong and, to the extent that they result in international governance, it is always appropriate to ask how such governance affects participants' interests. The degree of bias may make a considerable difference in a regime's durability, effectiveness, and mode of transformation. "Fairer" regimes are likely to last longer, as are those that call for side payments to disadvantaged participants. . . .

In the next two sections, we use the regime framework to discuss international relations in two contrasting issue-areas, 19th century colonialism and mid-twentieth century food affairs. Readers will recognize that these two regimes differ significantly along each of the four analytical dimensions we have elaborated. The colonial regime was diffuse, largely informal, subject to revolutionary transformation, and distinctly biased in distributing rewards. By contrast, the food regime is more specific, more formalized, probably in the process of evolutionary transformation, and more generally rewarding to most participants. Our primary intention is to highlight and clarify our theoretical definitions and the variables we have identified as useful for comparative analysis. Conclusions will push toward generalizations concerning regime outcomes and patterns of stability and change.

Colonialism, 1870–1914

Historians identify the years 1870 to 1914 as the heyday of European colonial expansion. Our analysis reveals that during this period the international relations of the imperial powers were regulated by a regime that prescribed certain modes of behavior for metropolitan countries vis-à-vis each other and toward their respective colonial subjects. Save for the United States, which entered the colonial game rather late (and Japan, which entered later and never participated in the normative consensus until after it had come under challenge), all of the colonial powers were European. The "regime managers" by 1870 were the governments of major states, where ministries and ministers made the rules of the colonial game and diplomats, soldiers, businessmen, and settlers played accordingly. In addition, a variety of subnational actors, including nebulous "publics" such as church societies, militarist lobbies, trade unions, and bankers, held opinions on issues of foreign policy and in some countries exercised substantial influence over the formulation of colonial policy.

The international relations of colonialism were evident in distinctive patterns of political and economic transactions and interactions. Flows of trade and money were typically "imperial" in the sense implied by Hobson or Lenin: extracted raw materials flowed from colonies to metropolis, light manufactures flowed back, investment capital flowed outward from European centers, and profits and returns flowed back.

But much more important than the characteristic transaction flows of colonialism were the interaction patterns in relations among imperial powers and between them and their respective colonies. There was a pronounced competitiveness among metropoles as each country sought to establish, protect, and expand its colonial domains against rivals. Yet there was also a sense of limitation or constraint in major-power relations, a notion of imperial equity, evidenced in periodic diplomatic conferences summoned to sort out colonial issues by restraining the expansiveness of some and compensating others for their losses. Constraint and equity were also reflected in doctrines like "spheres of influence" and "open doors," which endorsed the notion that sharing and subdivision were in order. . . . In inter-imperial relations, then, there were distinct elements of international management over selected parts of the non-European world. This management rested upon implicit codes for managing colonies, rationales like "civilizing mission," which were given credence, and growing willingness to agree on imperial borders by diplomatic conferences.

With regard to relations between the metropolitan powers and subject peoples, little equity prevailed. Commands, directives, and demands flowed from colonial ministries to colonial officers and then either to compliant local functionaries or directly to subjects. Deference and compliance flowed back. Defiance usually brought coercive sanctions, with success largely guaranteed by the technological superiority of European arms. . . .

Norms of the Colonial Regime

The legitimacy in colonization was founded upon consensus in a number of norms that the governments of the major powers recognized and accepted. These subjective foundations of the international regime may be treated under six headings.

a. The Bifurcation of Civilization. Looking from the metropolitan capitals outward, the world was perceived as divided into two classes of states and peoples, civilized and uncivilized. Europe and northern North America occupied the civilized category, and all other areas were beyond the pale, save perhaps other "white-settled" dominions. . . . From this, it followed politically that inequality was an appropriate principle of international organization and that standards and modes of behavior displayed toward other international actors depended upon which category those others fell into. . . .

b. The Acceptability of Alien Rule. The zenith of European imperialism occurred before the principle of national self-determination became a tenet of world politics, and indeed before Europe itself had largely settled into the pattern of "one nation, one state". . . . Thus, the imposition of foreign rule and the superimposition of white elites on indigenous elites were approved as right and proper, especially when such behavior was also perceived as "civilizing" or "christianizing."

c. The Propriety of Accumulating Domain. During the period 1870 to 1914 states' positions in the international status hierarchy were determined in considerable measure by expanses of territory (or numbers of inhabitants) under respective national jurisdictions. Domain was the key to prestige, prestige was an important ingredient in power, and power was the wherewithal to pursue a promising national destiny. The expansion of domain was therefore accepted by the European powers as a legitimate goal of imperial foreign policy and, indeed, reluctance to pursue such policies was considered unorthodox; it raised questions about the according of status. . . .

d. The Importance of Balancing Power. Intra-European relations in the late 19th century were stabilized by principles of a multipolar balance of power (even though the bipolarization that would harden by the eve of World War I was already in evidence). There was a widespread recognition of the efficacy of the balance of power and a general consensus among foreign offices that it should be preserved and perfected. This principle also justified colonial expansion and it further supported the norm of compensation. As a matter of right all colonial governments expected compensation for adjustments in the boundaries of colonial empires. . . .

e. Legitimacy in Neomercantilism. Economic exclusivity was a norm of colonialism since, as we have noted, colonies were considered to be zones of economic exploitation. Hence metropolitan powers endorsed their rights to regulate the internal development and external commerce of their colonies for the benefit of the home country, and, when appropriate or necessary, to close their colonial regions to extra-empire transactions. . . .

f. Noninterference in Others' Colonial Administration. As colonial domains were considered to lie under the sovereign jurisdiction of metropolitan governments, external interference in "domestic" affairs was not countenanced. The colonial powers could, and did, chip away at each others' domains via strategic diplomacy and occasional military skirmishes. But seldom did any one power question the internal administration of another's colonies. . . .

It is easy to see how these various tenets of the colonial regime affected international behavior. They abetted behavior directed toward establishing relationships of dominance and subordination, rationalized conquest and whatever brutalities it might involve, justified subjugation and exploitation, impelled a continuing major-power diplomacy concerning colonial matters, and necessitated periodic conferences and continuing bureaucratic-level communication. Such communication was aimed at limiting overexpansiveness, providing compensation, and maintaining the balance of power, and it had the effect of insulating empires from extra-imperial scrutiny and intervention. In this normative setting, colonization was deemed right and legitimate. It flourished.

The fundamental principles of the colonial regime were all challenged, even in their heyday, and eventually undermined during the years after 1920. By the 1970s dominance-subordination was considered an illegitimate mode of international relations, alien rule had become anathema, economic exploitation was condemned and attacked, territorial compensation was considered diplomatically ludicrous, and the internal affairs of empire (of which only small remnants remained) became matters of continuing international public disclosure and debate in the United Nations and elsewhere. . . .

Why did the regime change? First, and obviously, the power structure of the international system change; western European power was drained in two world wars; the United States and the Soviet Union rose to fill the power vacuum; new elites had come to power in both the United States and Russia after World War I and their preferences were distinctly anticolonial (though for ideologically different reasons). After World War II, new power emerged to buttress new principles and to support new institutions like the United Nations, where anticolonialism, promoted by the Soviet Union and acquiesced in by the United States, was taken up by smaller countries and proclaimed by excolonial states, whose ranks swelled yearly. A new global consensus was formed in the General Assembly under pressure from the Committee of Twenty-Four, and this held the tenets of the new anticolonial regime that prevails at present.

Some Analytic Characteristics of the Colonial Regime

The international regime that governed turn-of-the-century European colonialism was obviously diffuse, both geographically and functionally. Its tenets pertained to relations among metropolitan. countries, and to relations between them and their subjects. Whatever the substance of relations among metropoles, principles of exclusivity, compensation, and power balancing applied. In metropolitan-colonial relations of whatever substance principles prescribing dominance-subordination and abetting exploitation applied. To the extent that there was also geographically or functionally specific sub-regimes operative during the imperial era, such as American hegemonism in the Caribbean, the antislavery system, or intracolonial trade, they tended to embody as given the main tenets of the colonial regime. Interestingly, the colonial regime itself embodied some of the more general principles of 19th century international relations, as for example the central and explicit importance of power balancing, and the linkage between international stature and control over "domain." This suggests the hierarchical interrelationship of *superstructural* and *substructural* regimes discussed earlier.

Managing the colonial regime was a pluralistic exercise conducted largely by mutual monitoring and self-regulation practiced in national capitals. The regime was therefore, by and large, informal; there were few codified rules and no permanent organizations. . . . Formalization would have amounted to a spelling out of the rules, as for example those necessitating compensation, and would have jeopardized major-power relations by calling attention to constraints that rival governments could not admit to in public. The less said formally, the better, and the more durable the regime.

While our description of the colonial regime only hints at its transformation in the middle of the 20th century, the change was obviously of the revolutionary variety. There was little changing of minds or goals on the part of the colonial powers (save perhaps for the United States, whose government began to seek decolonialization almost as soon as the Pacific territories were annexed). Instead, counterregime norms took form in the European colonies in the 1920s and 1930s as nationalist elites emerged and movements were organized. The Russian Revolution created a formally anti-imperialistic state, thus breaking the European consensus that supported the principles of colonialism and modestly transferring power from the forces of imperialism to its challengers. Two world wars in the first half of the 20th century eclipsed European power and with it the capacity to retain great empires. After World War II the United States became aggressively anti-imperialistic for a time, thus shifting more power away from the supporters of the colonial regime. With the onset of the Cold War the United States subdued its anticolonialism in the interest of western unity (but Washington never admitted the legitimacy of empires). Meanwhile, counterregime norms prescribing decolonization had been legitimized and institutionalized by the United Nations General Assembly and its subsidiary bodies in the early 1960s. As the power to preserve the old regime waned, the power to replace it expanded. Personalities changed, norms changed, and power changed. As a result an international regime was discredited, eliminated, and replaced. The transformation was nothing less than a comprehensive change in the principles by which governments conducted their international relations.

Food, 1949–1980

Food constitutes a functionally rather specific regime, at least in comparison with diffuse regimes such as colonialism. Nonetheless, it conditions diverse policies and activity. Food trade, food aid, and international financing for rural development and agricultural research, for example, are all affected by the principles and norms of the international food regime. In contrast to the colonial regime, the food regime is more formal. Several organizations shape and spread regime norms and rules, and many rules are explicit and codified. . . .

Many of the regime's principles and norms are codified in treaties, agreements, and conventions such as the FAO [Food and Agricultural Organization] Charter, the International Grains Agreement, and the Food Aid Convention. The norms of the food regime are biased to favor developed and grain-trading countries, which have long enjoyed special weight in the IWC and FAO forums. Still, in contrast to the colonial regime, most participants in the food regime benefit to some extent from their compliance with norms. The regime is now in transition though, again in contrast to the colonial regime, change is taking place in evolutionary fashion.

For illustrative purposes we will focus on wheat as a key commodity in the international food regime, mainly because the international economics and politics of wheat have been thoroughly researched and we can therefore discuss regime influences with some confidence.

The national actors dominating the international wheat market since World War II have been the United States and Canada. In 1934–38 these North American countries supplied 20 percent of the wheat, coarse grain, and rice traded, while in 1979 they supplied 70 percent. These countries also held very large surpluses until 1972 (a byproduct of their domestic agricultural politics). Their common interests led them to operate as an informal duopoly. Together they controlled and stabilized international prices for two decades, though at the cost of allowing the price of internationally traded wheat to decline in constant terms by nearly one-half between 1950 and 1963–69. Other important actors in the food regime, as reflected by their participation in the wheat sector, include 1) major producers and consumers such as members of the European Communities (EEC), eastern European countries, and the Soviet Union; 2) other principal exporters such as Australia and Argentina; 3) poor importers such as China, Bangladesh, and Egypt; and 4) various international bodies such as the World Food Council, the Committee on Surplus Disposal of the FAO, and the major grain-trading firms. Of course, in more concrete terms the participants in the food regime are not really states and organizations but individuals, an international managerial elite of government officials who are responsible for food and agricultural policy within countries, and for bargaining about food affairs in international forums. Their network usually includes executives from the trading firms, some scientific experts, and occasionally representatives from public-interest organizations. But its core is a cluster of agricultural and trade officers. . . .

Norms of the food regime

In regulating food affairs over the last several decades, regime managers have been able to find consensus on a number of norms. Some of these reflect the overarching principles or superstructure of the state system; others are more specifically aimed at regulating food transfers. Eight norms in particular tend either to be embodied in the charters of food institutions, or to be recognized as "standard operating procedures" by food managers.

Respect for a Free International Market. Most major participants in the food trade of the post-World War II era adhered to the belief that a properly functioning free market would be the most efficient allocator of globally traded foodstuffs (and agricultural inputs). . . . Communist countries did not accept this norm for Soviet bloc trade, but abided by it nonetheless in East–West food trade. Actual practice often deviated rather markedly from free-trade ideals, as the history of attempts at demand and supply controls testifies, but in deference to the regime norm these were either rationalized as means towards a free market or criticized for their unorthodox tenets.

National Absorption of Adjustments Imposed by International Markets. This derives from norms worked out in the more diffuse trade and state-system regimes. The relative price stability that prevailed in international grain markets during much of the postwar era can be accounted for in large measure by American and Canadian willingness to accumulate reserves in times of market surplus and to release them, commercially and concessionally, in times of tightness. Of course, these practices occurred largely for domestic reasons. Yet their was still the almost universal expectation that North Americans could and would hold reserves for the world and would manipulate them in the interest of market stability. Hence this major norm—that each group dependent on the market should bear, through its own policies, burdens created by large price swings—was made easier to maintain as long as North American reserves acted to prevent large price variations.

Qualified Acceptance of Extramarket Channels of Food Distribution. Food aid on a continuing basis and as an instrument of both national policy and international program became an accepted part of the postwar food regime. By 1954 it was institutionalized by national legislation in the United States and by international codes evolving through the FAO's Committee on Surplus Disposal. Yet in a system oriented toward free trade, participants' acquiescence in extramarket distribution could be obtained only on the stipulation that market distribution was to take precedence over extramarket distribution. Therefore, food aid was acceptable to American and foreign producers and exporters as long as aid did not dramatically reduce income from trade or distort market shares. Rules to this effect were explicitly codified in national and international law. . . .

Avoidance of Starvation. The accepted international obligation to prevent starvation is not peculiar to the postwar period; it derives from more remote times. There has been and remains a consensus that famines are extraordinary situations and that they should be met by extraordinary and charitable means.

The Free Flow of Scientific and Crop Information. Whereas most of the other norms of the international food regime (and, more specifically, the wheat regime) emerged during the postwar era largely because of American advocacy and practice, "free information" emerged in spite of U.S. misgivings. Freedom of information about the results of agricultural research was a notion nurtured by the FAO and welcomed by those seeking technology for development. With American acquiescence, especially after 1970, it became a norm of the food regime and has become nearly universal.

Low Priority for National Self-Reliance. Partly because the global food system of the past thirty years was perceived by most participants as one of relative abundance and partly because of international divisions of labor implicit in free-trade philosophies, national self-reliance in food was not a norm of the international food regime. Indeed, food dependence was encouraged and becoming dependent upon external suppliers was accepted as legitimate and responsible international behavior. . . . Measures that reduce dependence in rich importers, such as the Common Agricultural Policy of the EEC and the subsidization of domestic rice production in Japan, . . . conflict with this norm implicitly. Such policies, however, reflect domestic political pressures rather than explicit goals of food self-reliance; hence they do not directly conflict with the emphasis on food trade *per se.*

National Sovereignty and the Illegitimacy of External Penetration. The international food system of the last thirty years existed within the confines of the international political system, so that principles governing the latter necessarily conditioned norms of the food regime. Among them, the general acceptance of the principle of national sovereignty largely proscribed external interference or penetration into matters defined as "domestic" affairs. In practice this meant that food production, distribution, and consumption within countries, and the official policies that regulated them, remained beyond the legitimate reach of the international community; a "look the other way" ethic prevailed even in the face of officially perpetrated inhumanities in a number of countries. . . .

Low Concern About Chronic Hunger. That international transactions in food should be addressed to alleviating hunger and malnutrition, or that these concerns should take priority over other goals such as profit maximization, market stability or political gains, were notions somewhat alien to the international food regime of the postwar era. It was simply not a rule of international food diplomacy that hunger questions should be given high priority, or, in some instances, it was not considered appropriate that they should even be raised when there was a danger of embarrassing or insulting a friendly country by exposing malnutrition among its citizens. . . .

Regime Consequences

Some effects of the prevailing food regime upon the international food system during the postwar era are easily discernible. In setting and enforcing regime norms for commercial transactions, the United States worked out trading rules in conjunction with key importers and other exporters. . . . Communist countries remained peripheral participants in these arrangements. They worked out their own rules within COMECON, although they occasionally interacted with "western" food traders, playing by western rules when they did. World trade in foodstuffs attained unprecedented absolute levels, and North America became grain merchants to the world to an unprecedented degree. Through concessional transactions the major problems of oversupply and instability in the commercial markets were resolved. Surpluses were disposed of in ways that enhanced the prospects for subsequent growth of commercial trade by the major food suppliers. Especially with respect to grain trading, adherence to regime norms enhanced the wealth and power (i.e., market share and control) of major exporters, most notably farmers and trading firms in the United States. The nutritional well-being and general standard of living of fairly broad cross-sections of populations were also enhanced within major grain-importing countries. Adhering to regime norms, however, also encouraged interdependence among exporters and importers, an interdependence that, over time, limited the international autonomy and flexibility of both. With regard to concessional food flows, regime norms facilitated global humanitarianism and enhanced survival during shortfalls and famines. . . . On the other hand regime norms also contributed to huge gaps in living standards between richer and poorer countries; they helped to perpetuate large gaps between rich and poor within countries; and they failed to correct chronic nutritional inadequacies of poor people worldwide. By promoting transfers of certain types of production technology as well as foodstuffs, the food regime also contributed to the spread of more capital-intensive farming and specialized rather than self-reliant crop choices. Overall, the food regime reflected and probably reinforced the global political–economic status quo that prevailed from the late 1940s to the early 1970s. It was buttressed by, and in turn buttressed, the global power structure of American hegemony.

The period 1970–78 was one of substantial instability in markets and concern for food distribution, food insecurity, and malnutrition. . . . Production actually declined worldwide and the traded tonnage expanded dramatically, and wheat prices tripled in the two years between the summer of 1972 and 1974. Sufficient concern was aroused both in and beyond the circle of elite regime managers that a World Food Conference was held in November 1974, to institute a series of reforms in the regime. Three substantive major defects in the world food system, as well as many minor ones, were identified at the Conference. The first was inadequate food reserves to assure reasonable stability in markets and security for consumers; second, the use of food aid in ways that reflected low priority for the food problems of less developed countries; and third, inadequate and inappropriate investment flows with respect to food production capacity in food-deficit areas.

These defects arose because behavior according to regime norms, which previously had not led to a conflict between domestic and international interests, now did so. The stockpiles that had guaranteed international price stability (though not the food security of those unable to buy food) had not been created or maintained with the purpose of providing international stability. No norm had been institutionalized that prescribed reserves for international purposes. Reserves, held mainly by the United States and Canada, had been largely a function of political and economic responses to income demands of the politically significant farm populations in exporting countries. The norm held that adjustment to market conditions was a national responsibility. When reserves were no longer required for adjustment purposes in North America they were gladly, not cautiously, depleted. . . . Food aid was a mechanism to reduce such stocks and to promote new markets. In addition, the largest donor, the United States, allocated the bulk of its food aid on the basis of political rather than nutritional criteria and in direct proportion to the size of American stocks. . . . Dietary adequacy in poorer countries was not prescribed by the regime. Thus, Egypt, South Korea, Taiwan, Israel, and even Chile got aid, while near-famine occurred in Bangladesh. Finally, investment in food production, especially in poor countries, was low because it was not seen as attractive by the dominant philosophy of economic development— import-substituting industrialization. Nor was it relevant to the largest motivation that shaped private capital flows, namely, the search for cheap sources of supply. As noted, agricultural development was not prescribed by the regime. These considerations abetted the transfer of existing rather than new technologies, and leaned against investment in rural areas and in food crops for local consumption (as opposed to fibers or tropical products such as coffee, tea, and pineapples). . . .

In the seven years since the World Food Conference, has the regime changed? The answer is "marginally," and by evolution. First, there has come to be a greater emphasis on rural development. . . . New norms emphasizing food in development planning have been codified by a special conference on rural development held in Rome in 1979 and by continuing World Food Council resolutions. . . . A second change is that greater security for food-aid recipients has been assured. This results from a new Food Aid Convention, agreed to in March 1980, which raised the minimum aid donor pledges from four-and-half to eight million tons, and from the four-million-ton emergency international wheat reserve of the United States signed into law in January 1981. . . . A violation of older, nonintervention norms is reflected in the way this aid intervenes in the domestic food policy of aid recipients. Another norm change is reflected in increased programming of food aid according to nutritional rather than political criteria, which has occurred in the food aid programs of the WFP, Australia, Canada, and Europe.

These changes constitute the evolution of new norms that challenge the priority of market principles and give higher priority to chronic hunger, food security, and food self-reliance. . . .

In other respects the tenets of the old regime prevail, and priorities remain as they were in the early 1970s. National policies dominate international policies and "free market" mechanisms are still held to be ideal for the bulk of food allocations flowing in international channels. With respect to reserves to increase stability and security, progress has been limited. . . .

Food security remains tenuous. The internationally coordinated system of reserves for food security called for at the World Food Conference has not been created. . . .

All current trends suggest that food deficits will grow in a number of regions of the world, particularly in Asia and Africa. Furthermore, the rising cost of production, ecological deterioration, and the decline of subsistence agriculture all point to increasing vulnerability in the relationship between food supplies and needy customers. Regime changes since 1974 to cope with this problem of maldistribution are almost certainly inadequate.

We expect that the higher degree of formalization of the food regime and its large number of voluntary participants will lead to continuing and accelerated efforts to change substantially the norms of the food regime. These efforts will occur within the frameworks of public and private organizations—United Nations agencies, special forums and secretariats, centers, councils, committees, conferences, and companies. Setting the rules of the game will remain the prerogative of powerful national governments, especially those whose foodstuffs dominate in trade and aid flows, and in particular the United States. To the extent that there has been incremental change in the global food regime, pressures from formal international institutions have been helpful. . . .

Otherwise, the food regime does not look very different or function very differently from other regimes. Its principles have legitimized unequal distributions of food and unequal distributions of benefits from buying and selling food. Its norms have given authority to the powerful, both informally and in formal international bodies. Interestingly, though, there has been little formulation or articulation of revolutionary norms.

Conclusions

Six General Conclusions

Our two cases, colonialism and food, suggest some conclusions. They are hardly definitive or universal, but they might be subject to broader generalization and further refinement.

Without intending to be trivial let us first underline that *regimes exist*. In international relations there are revered principles, explicit and implicit norms, and written and unwritten rules, that are recognized by actors and that govern their behavior. Adherence to regimes may impose a modicum of order on international interactions and transactions. Our two case studies demonstrate that actors are guided by norms in diverse issue-areas. We would suggest that regimes exist in all areas of international relations, even those, such as major-power rivalry, that are traditionally looked upon as clear-cut examples of anarchy.

Second, taking regimes into account contributes to explaining international behavior by alerting students of international affairs to subjective and moral factors that they might otherwise overlook. Once this subjective dimension of international relations is included, explanations of international behavior can be pushed beyond factors such as goals, interests, and power. Our case study reveals that regimes mediate between goals, interests, and power on the one hand, and behavior on the other. Such normative mediation is most effective, and hence most theoretically significant, between two limiting sets of conditions. At one extreme, a regime may be an empty facade that rationalizes the rule of the powerful by elevating their preferences to the status of norms. Under such conditions a regime exists because subordinate actors recognize the rules and abide by them, but knowing this would not significantly improve upon our ability to explain behavior as all we would need to know are the identities of the powerful and their interests and goals. Under the colonial regime, for example, knowledge of norms contributed little to explaining the dominance of metropolis over colonies. Similarly, under the food regime, the knowledge that there were norms revering free markets does not contribute greatly to explaining major trends in the trading behavior of the major exporters. These actors pressed for free trade because it was in their interest. . . .

At the other extreme are conditions where regimes are determinative, where codified international law or morality is the primary guide to behavior, and where the separate goals, interests or capabilities of actors are inconsequential. Such conditions are extraordinarily rare in international relations. Where they prevail (in narrow, highly technical issue-areas like smallpox control or international posts and telegrams) consequent international behavior is analytically uninteresting. . . .

Between the limits of major-power hegemony and legal or moral order is a rather broad range of international relations where regimes mediate behavior largely by constraining unilateral adventurousness or obduracy. The case studies suggest conditions under which such normative mediation takes place. For example, it occurs in relations among powers of comparable capability, where the exertion of force cannot serve interests. Here, norms and rules tend to order oligarchies, establishing the terms of a stable and peaceful relationship, mediating and moderating conflict, and preserving collective status and prerogatives against outsiders. Relations among the colonial powers, for example, were obviously mediated by norms, and knowing this adds to our ability to explain behavior that had large consequences for colonial regions. Under the food regime, exporters' direct and indirect relations with each other were mediated by norms such as those proscribing concessional dealings until commercial markets were cleared. . . .

Regimes also mediate under conditions of diffused power, or under conditions where asymmetries in power are neutralized, as in one-state-one-vote international forums. Here, consensus about appropriate decision-making procedures and their legitimacy keeps pluralism from deteriorating into anarchy, and consensus about legitimate objectives makes policy possible.

Finally, regimes mediate during transitions of power. They tend to have inertia or functional autonomy and continue to influence behavior even though their norms have ceased either to reflect the preferences of powers or to be buttressed by their capabilities. This is one of the most fascinating and useful aspects of regime analysis, where compliance with norms explains why patterns of behavior continue long after reasoning in terms of power and interest suggests that they should have disappeared. . . .

Our third conclusion is that functionally specific and functionally diffuse regimes differ importantly with regard to the locus of management and the nature of managers. Functionally specific regimes such as the food regime are directed by technical specialists and middle-echelon administrators in participating governments. Such officials are recruited for their expertise and skills, traits that are well dispersed internationally. As a result, specific regimes tend to follow rather democratic procedures, at least as concerns policies pursued by managers. By contrast, functionally diffuse regimes such as the colonial regime are more often managed by diplomatic generalists and higher-level political officers. Not only does this suggest that diffuse regimes are likely to be much more highly politicized than specific ones, but also that conflicts which arise in the contexts of various regimes will be different. Resistance to issue linkage, for example, will be more common in specific regimes, where managers will variously seek to insulate (or, alternatively, expand) their jurisdictional domains. On the other hand, difficulties in enforcing norms, and greater deviance and regime challenges, are likely in diffuse regimes.

Fourth, international regimes are formalized in varying degrees. Our analysis suggests that degrees of formality tend to have relatively little to do with the effectiveness of regimes measured in terms of the probabilities of participants' compliance. With the two regimes we considered, one formal and one informal, both predictably and consistently constrained most participants' behavior over considerable periods of time. The colonial case suggests that some of the most effective regimes are those that are quite informal. This would seem to be true especially for regimes that regulate the general political behavior of major powers. . . .

While there may be few differences in the effectiveness of formal and informal regimes, our analyses suggest that "formalization" itself may be a dynamic factor. Regimes tend to become more formal over time, as with the colonial regime, where multilateral diplomatic conferences became increasingly important in the latter years of the imperial system; or with the food regime, where organizations, institutions, and rules seem now to be proliferating to fill a void in management created by American reluctance to provide informal leadership. We believe that regimes formalize over time because maintenance often comes in one way or another to require explicitness. As those rewarded by a regime's functioning become either accustomed to or dependent upon such benefits, they tend to formalize interaction patterns in order to perpetuate them. . . .

Fifth, effectiveness in terms of compliance with rules and procedures of any given regime depends largely upon the consensus or acquiescence of participants. Formal enforcement is extraordinary and coercive enforcement is rare despite its prevalence in relations between metropolis and colonies during the colonial era. Usually it is self-interest, broadly perceived, that motivates compliance.

Most participants in international regimes, whether they are advantaged or disadvantaged under the regime's normative biases, usually comply because compliance is calculated to be more rewarding or less costly than deviance. Saying this is perhaps pushing the obvious. But what is intriguing is how regime participants calculate their benefits and costs, and especially how they assign weights to perceived "moral" benefits of acting in accord with norms, or perceived "moral" costs of acting against them. . . .

Sixth and finally, our comparative case studies of regimes suggest that regime change is closely linked to two classical political concepts—power and interest. Most regime change results from changes in the structure of international power. For diffuse regimes, the relevant power structure is the global political-strategic balance, as was the case with the colonial regime, which began to change when major powers such as Russia (the Soviet Union) and the United States defected from the normative consensus. On the other hand, for more functionally specific regimes, relevant power also must include command over specific resources within particular issue-areas, as with the oil companies during the 1930s and the oil states in the 1970s, and the food-supplying states in the food regime. Of course, principles such as sovereignty may extend from the diffuse state system to affect or be part of the features of these specific regimes as well.

Revolutionary change is the more frequent pattern of regime change, and such change most often comes after changes in the structure of power. On the other hand, regime change via cognitive learning and the recasting of goals among dominant elites also occurs. This evolutionary change seems less frequent than revolutionary change, perhaps because major wars, from the Thirty Years War to World War II, have preceded and been instrumental in regime change.

Regime change without significant changes in power structure occurs when leading elites seek to preserve their status and their control of the regime by eliminating "dysfunctional" behavior, either in the substantive performance or in the decision procedures of a regime. This results when learning and technology foster new or changed goals. Changes in interests and goals have arisen from expanding knowledge of the world and its environmental exigencies. New understanding and capability with respect to disease, food technology, and air travel are important instances of regime change and even regime creation. The norm that no one should be hungry is not accepted by the current food regime, but it has sparked major efforts at regime change, including the creation of international reserves and external aid to increase food production in areas of the world that are most chronically malnourished. Unfortunately, it is only rarely the case that controlling elites—especially the fragmented and oligarchic elites of the international system—learn enough in sufficient time to change from within.

Transforming International Regimes:
What the Third World Wants and Why

STEPHEN D. KRASNER

Introduction

Developing countries have pursued many objectives in the international system. Some objectives have been purely pragmatic, designed to enhance
immediate economic well-being. However, the most publicized aspects of
North–South relations, global bargaining over the restructuring of international regimes, cannot be understood in strictly economic or instrumental
terms. By basically changing principles, norms, rules, and procedures that
affect the movement of goods and factors in the world economy, the Third
World can enhance not only its economic well-being but also its political
control. The emphasis the South has given to fundamental regime change is
a manifestation of four basic factors: the international weakness of virtually
all developing countries; the domestic weakness of virtually all developing
countries; the systemic opportunities offered by the international institutions
which were created by a hegemonic power now in decline; and the pervasive
acceptance of a belief system embodying a dependency orientation.

    At the international level all states are accorded formal equality as sover-
eigns. The underlying power capabilities of states establish no presumptive
differentiation with regard to certain basic rights, especially sole legitimate
authority within a given geographic area. At the same time, the present
international system is characterized by an unprecedented differentiation in
underlying power capabilities between large and small states. Never have

Reprinted from *International Studies Quarterly,* Vol. 25, No. 1, March 1981, "Transforming
International Regimes: What the Third World Wants and Why?" Stephen D. Krasner, by permission of *International Studies Quarterly.* © 1981.

states with such wildly variant national power resources coexisted as formal equals. Very weak states can rarely hope to influence international behavior solely through the utilization of their national power capabilities. For them, regime restructuring is an attractive foreign policy strategy, because it offers a level of control over states with much larger resources that could never be accomplished through normal statecraft grounded in dyadic interactions.

The rigidity and weakness of domestic economic and political structures in developing countries is a second factor that has made basic regime change important for the Third World. With the exception of a small number of countries, the economies of the Third World are dominated by agricultural and primary sectors with low levels of factor mobility. Vulnerability is high because it is difficult to adjust to external changes. Political systems are also weak; the state cannot manipulate those resources that might lessen the impact of pressures emanating from the international environment. International regimes can limit external vacillations or automatically provide resources to compensate for deleterious systemic changes.

The third element accounting for the prominence of a basic regime change strategy is the set of opportunities offered by the character of post-World War II international organizations. These organizations have offered opportunities that made Third World programs more feasible and effective. The Third World has been able to turn institutions against their creators. Such developments are likely to afflict any set of regimes created by a hegemonic power. This power establishes institutions to legitimate its preferred norms and principles, but legitimation can only be effective if the institutions are given independence and autonomy. This autonomy can then be used by weak states to turn the institutions to purposes and principles disdained by the hegemonic power.

Affecting both domestic incentives and international opportunities in the Third World's quest for a new international economic order has been a belief system associated with theories of dependency. This intellectual orientation has been a critical factor, accounting not only for some of the Third World's success, but also for its extraordinary unity on questions associated with regime transformation. Even economically successful developing countries with flexible domestic structures and conservative political regimes have not broken with the Group of 77. In an atmosphere pervaded by *dependencia* perspectives, such a break could undermine a regime's position with domestic elements. No Third World state openly endorses the norms and principles of international liberalism, even if some of them adopt its rules and procedures. The ideological hegemony enjoyed by the United States at the conclusion of World War II has totally collapsed, and the alternative world view presented by dependency analyses has forged the South into a unified bloc on questions related to fundamental regime change.

The Variety of Third World Goals

The emphasis in this essay on weakness, vulnerability, and the quest for control is not meant to imply the LDCs are uninterested in purely economic objectives. Third World states have pursued a wide variety of goals. These include economic growth, international political equality, influence in international decision-making arenas, autonomy and independence, the preservation of territorial integrity from external invasion or internal fragmentation, the dissemination of new world views at the global level, and the maintenance of regime stability. They have used a wide variety of tactics to promote these objectives, including commodity organizations, regional coalitions, universal coalitions, alliances with major powers, local wars to manipulate major powers, irregular violence, bilateral economic arrangements, regulation of multinational corporations, nationalization of foreign holdings, foreign exchange manipulation, and international loans.

This essay does not review all aspects of Third World behavior. It concentrates on an area where political objectives associated with control have been highly salient—Third World efforts to enhance power through the transformation and construction of international regimes. By building or altering international institutions, rules, principles, and norms, weaker countries can both ameliorate the vulnerability imposed by their lack of national material-power capabilities and their weak domestic political structures, and increase resource flows.

Third World political behavior, like all political behavior, can be divided into two categories: relational power behavior which accepts existing regimes, and meta-power behavior which attempts to alter regimes. Relational power refers to the ability to change outcomes or affect the behavior of others in the course of explicit political decision-making processes. Meta-power is the capacity to structure the environment within which decisions are made. This structuring can involve the manipulation of institutional arrangements, norms, and values. Relational power behavior accepts the existing rules of the game; meta-power behavior attempts to alter those rules.

Outcomes can be changed both by altering the resources available to individual actors and by changing the regimes that condition action. Changing the outcome of struggles fought with relational power requires changing actor capability. However, such changes do not necessarily imply an alteration in meta-power. A state may prevail more frequently in disputes with other international actors by enhancing its national power capabilities without altering the institutional structures, norms, and rules that condition such disputes.

Outcomes can also be changed by changing regimes. Meta-power behavior is designed to do this. Successfully implemented, it implies a change in relational power as well. Individuals who win at poker may lose at bridge; political parties that secure seats under a proportional representation system might be excluded by single-seat districts; states that secure greater revenue from cartelized exports might be poorer if the price of their product was dictated by those with the greatest military capability. An actor capable of changing the game from poker to bridge, from proportional representation to single-seat, from economic to military capability, has exercised meta-power. Actors may seek to enhance their relational power by enhancing their own national capabilities, or they may attempt to secure more favorable outcomes by pursuing a meta-power strategy designed to change regimes.

Most studies of international politics have implicitly emphasized relational power because they deal with war and the use of force. In this arena, meta-power considerations are of limited import because institutional restraints, norms, and rules are weak. . . . War outcomes are determined by the relative national material capabilities for the actors involved: what resources are nominally under the jurisdiction of the state and how well the state is able to mobilize and efficiently deploy these resources.

In issue areas other than the use of force, however, regimes have been more salient. Wars involve relational power strategies based on national power capabilities; nonbelligerent issue areas are susceptible to meta-power strategies designed to alter regimes.

Third World states are interested in employing both relational power and meta-power. Proposals for regime change, voiced by the less developed countries, reflect an effort to exercise meta-power. The objective of these proposals, of which the program associated with the New International Economic Orders (NIEO) is the most recent and salient, is to transform the basic institutional structures, norms, principles, and rules that condition the international movement of goods, services, capital, labor, and technology. Such transformation is particularly attractive because the ability of Third World states to accomplish their objective solely through the exercise of relational power is limited by the exiguity of their national material-power capabilities. These alone could not resolve the vulnerability problems of poorer states. Developing countries have sought to alter regimes in a variety of issue areas. They have attempted to create new institutional structures or to change patterns of influence, particularly voting allocations, in existing structures. They have sought to establish new international norms. And they have tried to change rules. Many of these quests have been successful.

Third World Motivations

Third World demands for regime restructuring cannot be seen in any simple way as a reflection of economic failure. During the postwar period the overall rate of growth of developing areas has been faster than that of industrialized countries. Trade patterns have become more diverse with regard to partners and commodities. Indicators of social well-being, including life expectancy, infant mortality, and literacy, have dramatically increased in many areas. The economic performance of the South during the postwar period has been better than that of the industrialized countries during the nineteenth century.

However, the South continues to suffer from an enormous gap in power capabilities at the international level and from social rigidity and political weakness at the domestic level. Creating new regimes that reflect Southern preferences is one way to deal with these structural weaknesses.

International Structures

There have always been small states in the modern international system. Before the industrial revolution, however, there was little variation in levels of economic development. With regard to per capita income, the richest country was only about twice as well off as the poorest at the beginning of the nineteenth century. Now, the richest countries are 80 to 100 times better off than the poorest. The combination of small size and underdevelopment has left many Third World states in an unprecedentedly weak position.

In 1830, the ratio of the GNPs of the largest state, Russia, to the smallest state for which figures are available, Denmark, was 41:1. In 1970, the ratio of the national incomes of the largest state, the United States, and the smallest, the Maldives, was 97,627:1. By 1970, 34% of the states in the international system had national incomes that were less than one thousandth that of the United States and 72% had national incomes less than one hundredth of the U.S. figure. These are staggering disparities. In 1970, the Third World as a whole accounted for only 11% of world GNP.

With the exception of China, there is no Third World country that can lay claim to great power status. However, the GNP of the United States is 5.64 times larger than China's. Countries proffered as regional hegemonic powers do not have impressive national power capabilities. In the mid-1970s, the GNPs of India and Brazil (the two largest in the Third World after China) were about the same as those of Spain and Poland; Iran's (and this before the Khomeini regime) rivalled Belgium's; Saudi Arabia and Nigeria had GNPs about equal to those of Denmark and Finland.

Using GNP figures as a measure of power capability has the advantages of easy comparability and accessibility; however, it also has the disadvantage of obscuring potential variations in power capabilities across different issue areas. Yet, even at a disaggregated level, there is little evidence that Third World countries can act effectively by utilizing only their national material resources. In the area of raw materials cartelization efforts have failed—with the exception of oil—although coffee exporting states have had sporadic success in pushing up prices by buying in London and New York, and copper producers in withholding stock from the market. The fundamental problem for the exporters of primary commodities is that there is a high temptation to cheat on any cartel scheme, because the marginal rewards of additional revenues for Third World governments strapped for resources are very high. With regard to trade in manufactures, Third World exporters depend far more on Northern markets than industrialized countries do on manufactured goods from the South. Northern countries have import competing industries capable of producing the same products, while the South does not have alternative markets. With regard to bank lending, large Third World debtors, especially Mexico and Brazil, have secured some leverage through the consequences of default. While this has given them continued access to credit markets, it has not enabled them to alter the basic nature of credit relations or to keep interest rates down. Smaller debtors carrying heavy burdens are rolling over their old debt but having difficulty securing new loans.

There are two major exceptions to these comments about Third World national power capabilities in specific issue areas. The first is OPEC, where the combination of excess financial resources and inelastic demand has enabled Third World countries to raise prices eightfold in nominal terms over the last seven years. The second is national control of multinational corporations. Many developing countries have excluded MNCs from certain sectors, nationalized or unilaterally altered the concessions of petroleum and hard mineral corporations and limited the ownership share of foreign nationals either generally or in specific industries. Control over access to their territory has been an important source of leverage for LDCs. Host-country nationals have also learned about market access and technology, which has given them more bargaining power. However, the pressure that can be exercised by host countries is limited by the ability of firms to relocate in more hospitable countries.

Aside from oil and domestic regulation of MNCs, few Third World states have any ability to alter their international environment solely through the use of national material-power capabilities. Their small size and limited resources, even in specific issue areas, is the first condition that has led them to attempt the fundamental alteration of international regimes. Conventional statecraft based upon national material attributes is unlikely to reduce vulnerabilities. A meta-political strategy designed to alter rules, norms, and institutions offers an attractive alternative, if only be default.

Domestic Structures

The second condition that has driven Third World states to attempt a transformation of international regimes is the weakness of their own domestic societies and political systems. The international weakness of most developing states, as indicated by their small aggregate output in comparison with that of industrialized states, suggests that they cannot directly influence the international system. It also suggests that they will be subject to external forces that they cannot change. Small states are usually more heavily involved in the world economy. In 1973, trade (exports plus imports) was equal to 37% of GNP for developing countries, 29% for industrialized countries. In the same year, 48 out of 87 LDCs had trade proportions greater than 50%.

Although small states, as a rule, are more heavily involved in the world economy, state size does not determine internal capacity to modulate the pressures emanating from an uncertain international environment. A small, adaptable state could adjust to many regime structures. Such a state could accept its lack of influence at the international level but remain confident of its ability to deal with environmental disturbances over a wide range of international rules, norms, and institutions.

The ability to cope with environmental disturbances is a function of the mobility, flexibility, and diversity of a country's resources. A country with highly mobile, flexible, and diverse factors can absorb external shocks. It can adjust its pattern of production, imports, and exports to maximize its economic returns under different environmental conditions. Adjustments might be directed by the state or the private sector. The first alternative requires a strong political system, one in which the state is capable of resisting pressures from domestic groups, formulating a coherent strategy, and changing social and economic structures. The second alternative requires a well-developed private market with high levels of communications and information.

Social Rigidity

At early stages of development, countries lack the capacity to absorb and adjust to external shocks. This incapacity is produced by rigidities inherent in traditional structures. First, a status society is based upon exclusive corporate groups, which lock individuals within a rigid structure. Second, social action is determined by personal rather than impersonal norms. Different individuals are treated in different ways because of ascriptive characteristics. Third, the division of labor in the society is based on assignment to specific ascriptive groups. An individual's economic activity is permanently established by his group membership. Fourth, the ontology of the society stresses the concrete and discrete. General principles that can be applied to a wide range of situations are eschewed. Fifth, the world is seen composed of "concrete and discrete elements—that is, indivisible units—economic, social, cultural, and political resources are seen as being finite and immobile rather than expanding and flexible." By contrast, modern societies are market rather than status-based. Interactions are governed by impersonal norms of action. The individual and the nuclear family, rather than the corporate group, are the building blocks of the society.

Modern societies are less vulnerable to external changes because their factors are more mobile. Better-trained workers can perform a wider variety of tasks. More-developed capital markets can more readily reallocate investment resources. It is easier for an industrial worker to move from one factory to another than for a peasant to shift from one crop to another, much less move from agrarian to industrial employment. . . .

The transition from traditional to modern society is taking place in the Third World, but it is a slow and difficult process. It is not unidirectional or irreversible, as events in Iran and Cambodia demonstrate. Most developing countries are still in the early phase of the transition from a traditional to a modern economy which occurs at per capital income levels from $200 to $600 (in 1976 dollars). In this phase, societies are vulnerable to external shock. Most employment is still in agriculture. Cross-national data indicate that, on average, industrial output does not exceed agricultural output until per capita incomes of $800 are reached, and that industrial employment does not exceed agricultural employment until per capita income is $1,600.

 . . . . Most Third World countries have not moved very far along the path from tradition to modernity. While there is considerable variation among the countries of the Third World, a very sharp cleavage still exists between industrialized and developing countries.

Political Weakness

The rigidity of the social and economic structure in developing countries is reflected in the political system. Most central political institutions in the Third World are weak. The state is often treated as but one more compartmentalized unit. Its ability to extract resources from the society is limited. Efforts to combine diverse social and material units are likely to be frustrated by the compartmentalized nature of the society. Economic activity that takes place outside the market cannot be effectively tapped by the government. Often the state is unable to resist pressure from powerful society groups. Low levels of skill and education make it difficult to formulate effective economic policies. Under conditions of social mobilization and low levels of political institutionalization, the likely outcome is political decay rather than political development. The state is rarely able to adjust domestic structures in ways that would lessen the deleterious consequences of external changes.

Tax Structures

Tax structures offer the opportunity to illustrate differences between the political capabilities of industrialized and developing countries. Tax collection is generally a good indicator of the ability of the state to extract resources from its own society. Developing countries collect a smaller proportion of their GNPs than industrialized states and rely more heavily upon trade taxes; the level of state revenue is, therefore, more subject to international economic vicissitudes.

The NICs vs. OPEC

There is one major exception to these generalizations about weak political and rigid social structures in Third World countries. The newly industrializing countries, or NICs, have been able to adjust effectively to the international environment. Singapore, Hong Kong, Taiwan, South Korea, and Brazil have adopted aggressive export-oriented strategies. In Hong Kong the private market has acted effectively in a laissez-faire situation. In the other NICs the government has been more active. In Korea and Brazil for instance, the state explicitly decided to promote export-oriented growth and move away from protectionism in the early 1960s. Despite domestic pressure, both were able to maintain lower effective exchange rates, a precondition for international export competitiveness. Through the 1970s, the NICs were able to adjust to restrictions imposed by industrialized nations by developing new product lines and diversifying their exports. . . .

The NICs are one of the two groups of dramatic success stories with regard to economic growth, or at least transfers, in the postwar period. The other is oil-exporting states. If purely economic considerations are used to explain the behavior of developing countries, the difference in foreign policy orientations of countries in these two groups is difficult to understand. Both the NICs and the OPEC countries have dramatically benefited from the present system. While none of the NICs have taken a leading role in the South's efforts to restructure international regime, a number of OPEC countries have been at the forefront of the Third World movement. Algeria and Venezuela have taken leading roles in the Group of 77. Iraq, Libya, and now Iran are hardly devotees of the existing global order. . . .

Domestic structural weakness, a manifestation of traditional social norms, and political underdevelopment, together become a . . . factor that makes international regime transformation attractive for almost all Third World countries. The external environment is inherently threatening even in the absence of any direct effort by more powerful states to exercise leverage. International regimes controlled by developing countries can mitigate the exposure of developing areas to systemically generated changes. They offer some control in a situation where the lack of domestic adjustment capacity precludes effective cushioning against external shocks.

Systemic Opportunities

Demands for regime restructuring have occupied a dominant place in North–South relations, not simply because this approach could compensate for the international and domestic weakness of Third World states, but because the postwar system offered developing countries a setting in which to pursue this strategy: The prominence given to meta-political goals has been a function of opportunities as well as needs. The postwar liberal regime, especially the importance that it accorded to international organizations, provided the Third World with forums in which to press their demands.

The relationship between regimes and the underlying power capabilities of states can be assessed with regard to two issues. First, does the regime reflect the preferences of states weighted by their power capabilities? The characteristics of the regime may be identical with or diverge from the preferences of the most powerful states. Second, what effect does the regime have on relative underlying power capabilities? There are three possibilities: the regime may have no impact; it may reinforce the existing distribution; or, it may undermine the existing distribution. These possibilities are summarized in the following table.

The four most likely alternatives are labelled in Table 1. First, the power structure and regime may be congruent: The characteristics of the regime reflect the preferences of individual states (weighted by their national power capabilities) and do not affect capabilities. Second, there may be incongruence: The characteristics of the regime do not correspond with the prefer-

Table 1 Impact of Regime on Underlying National Power

State Preferences Weighted by National Power Capabilities and Regime Characteristics	Neutral	Reinforces Existing Power Distribution	Undermines Existing Power Distribution
Identical	Congruence	Dynamic Stability	
Divergent	Incongruence		Dynamic Instability

ences of individual states weighted by their national power capabilities, but the regime does not alter relative capabilities. Third, there may be dynamic stability between the power structure and the regime: The characteristics of the regime conform to the preferences of the strong and reinforce the existing power distribution. Fourth, there may be dynamic instability between the regime and the international structure: The characteristics of the regime do not correspond with the preferences of the strong and tend to undermine the position of more powerful states.

Developments in the postwar system suggest that during periods of hegemonic decline, there is a propensity to move from congruence toward incongruence or even dynamic instability. For hegemonic powers there is a paradox, perhaps an inevitable dialectic, involved in the creation of international regimes—including international organizations. Regimes that the hegemonic power initially creates to serve its own interests can be seized and restructured by other actors in the system.

Fully hegemonic powers are likely to establish a regime that is congruent rather than dynamically stable. Their national material dominance is so complete that they can ignore the impact of regimes on their relative power capabilities. They are likely to seek broad milieu goals. In contrast, a normal power is likely to attempt to create a dynamically stable regime. Since it lacks the slack resources possessed by hegemonic states, the normal power must be primarily concerned with enhancing its national power capabilities.

However, a situation of congruence resulting from regime creation by a hegemonic state is likely to be transformed over time into one of incongruence or even dynamic instability. This pattern of change can be traced to several factors. First, from the outset the hegemonic power is compelled to provide institutional structures with autonomy. The regime's purpose is to legitimate hegemonic preferences. Legitimacy cannot be promoted if the regime is perceived as merely an appendage of the hegemonic state. In the present international system, the primary source of legitimacy for rules and norms is their voluntary acceptance by individual states. This practice ultimately rests upon the concept of sovereignty, the dominant political principle of the modern era. The sovereign territorial state is the only actor whose authority is fully accepted in the international community. Sovereignty is indivisible, and in principle, all sovereign states are equal. There is, then, a prima facie case for distributing votes equally in international organizations.

When this norm is violated, as in the case of international financial institutions, the legitimating capacity of these organizations is weakened. Thus, there is a strong propensity for a modern hegemonic state to accept international organizations where voting power is equally distributed among all members. However, an equal division of votes opens the opportunity for weaker states to enhance their influence and control within these institutions.

A second endogenous factor that prompts change in the relationship between regimes and power structure is the independent inertia that can develop within international organizations. . . . This process of independent growth is again facilitated by the behavior of a hegemonic power during periods of regime creation. For institutions to legitimate a new set of rules and norms, they must be given autonomy from the dominant power. Objective criteria must be established for the selection of staff. Goals must be stated in general terms. Once this is done, the organization can fend for itself. It may generate its own ethos. It can respond to different opportunity structures. It can welcome new clients. As with individuals, organizational maturation is influenced by both genetic characteristics and environmental pressures. Over time the preferences of the organization and those of the declining hegemonic power may drift apart.

A third factor in the evolution of regimes from congruence to incongruence is the specific consequences for international organizations of a hegemonic power's decline. This is an exogenous consideration, which is generally independent of regime considerations. At the pinnacle of its power, the hegemonic state is prone to supply a disproportionate share of collective goods for the system, including international organization budgets. However, as it drifts toward the status of a normal power it will become more reluctant to do this, providing an incentive for international organizations to search for new sources of support. The hegemonic powers's influence on other states will also decline; they will be less likely to support its initiatives—within organizations as well as elsewhere. . . . Thus, international organizations can be seized by states whose national power capabilities are limited.

In the postwar period, the Third World has made international organizations a centerpiece of its demands for regime change. The South has succeeded in dominating the agendas of all major multifunctional universal organizations. The North has been compelled to respond rather than initiate. . . .

Debates and resolutions presented at international forums have altered norms, rules, and procedures in a variety of ways favored by developing countries. Various resolutions have endorsed 0.7% of GNP as a target for concessionary capital transfers from the North to the South. While this norm is more honored in the breach, it is still held up as a goal that has been accepted by the North as well as the South. . . .

In the area of trade, developing countries have used GATT to legitimate concessional treatment. During the 1960s, the industrialized countries agreed to institute a generalized system of preferences that would eliminate tariffs on some products from developing countries. The nontariff barrier codes and revisions to the GATT Articles of Agreement negotiated during the Tokyo

Round provide for special and differential treatment for developing countries, although more symmetrical behavior is expected as countries reach higher stages of development. These changes are a fundamental break with the two central norms of the postwar trading order: nondiscrimination and reciprocity. The South has enshrined new principles emphasizing development and equity, not just secured exceptions from the old liberal rules.

Even in the area of monetary affairs, that bastion of postwar conservatism, the South has had some success, at least within the regime's formal institutional manager, the IMF. While LDCs did not get an aid link with SDRs, they did get an allocation based on quotas. The industrialized nations had originally wanted virtually to exclude developing countries. The partial use of IMF gold sales to establish a Trust Fund (which makes loans to developing countries with few conditions at concessional interest rates) is a form of international taxation for aid. The Fund has begun to liberalize its conditions for stand-by agreements. In an international environment, in which the scope and growth of fund activities will depend in part on continuing willingness of developing countries to use its resources, the organization has moved to change its rules and procedures if not its basic principles.

In general the institutional structure has become more responsive to the South. By using its voting majority in the General Assembly, the South has been able to create new institutions, especially UNCTAD and UNIDO, which present its interests. Even in established forums, where votes are not equally divided, the South has changed voting power and decision-making procedures. Mutual veto voting arrangements for major decisions now prevail in all international financial institutions, including the Fund. In the Inter-American Development Bank, the largest of the regional lending institutions, and in the United Nations Development Program the Third World has a majority of votes. In the newest international financial institutions, the International Fund for Agricultural Development, votes are equally divided between OPEC countries, non-oil developing countries, and industrialized countries.

Thus, in a variety of issue areas the South has been able to alter principles, norms, rules, and procedures. It is difficult to imagine similar success in the absence of institutional structures that provided automatic access for developing countries. By taking advantage of the autonomy that the hegemonic power, the United States, was compelled to confer on international organizations during the period of regime formation at the conclusion of World War II, Third World countries have been able to alter regime characteristics during the period of American hegemonic decline. The relationship between underlying national power capabilities and regime characteristics has become increasingly incongruent.

Belief Systems

While vulnerabilities that arise from domestic and international weakness provide the impetus for Third World demands—and international organizations the opportunity to realize them—the form and unity of these goals have been shaped by the pervasive acceptance of dependency orientations. Most developing countries have explicitly accepted arguments that attribute their underdevelopment to the workings of the international economic system rather than the indigenous characteristics of their own societies. The belief system has been endorsed, not only by individual states, but by international organizations close to the Third World, such as UNCTAD and the UNDP, as well as by important groups with claims to speak for the North as well as the South. Individual states may reject dependency prescriptions in practice but even the most conservative lack a belief system to offer in its stead.

The dependency orientation serves important functions for Third World states both internationally and domestically. At the international level, dependency arguments have provided a unifying rationale for disparate Southern demands. Calls for special and differential treatment are justified by the contention that the South has been treated unjustly in the past. Existing norms and rules are rejected as inherently exploitative. A coherent intellectual orientation has been particularly important because of the strategy of using international organizations to promote meta-political goals. In such arenas the ability to define issues and control the agenda is critical. Such initiatives are facilitated by a widely shared and internally consistent analytic framework.

Dependency perspectives are also linked to domestic political conditions in Third World countries. Given the limitations on effective state action, foreign policy is an attractive way to build support. Prominence in universal coalitions can enhance a Third World leader's domestic position. Castigating the North can rally bureaucratic, military, and popular elements. The structure of international organizations affords Third World statesmen an opportunity to play on the world stage, a platform which they could not mount if they had to rely solely on the domestic power capabilities of their countries. . . .

Third World leaders who follow such a course must find ideological arguments that resonate with their domestic populations. The most accessible themes reject existing international regimes. For most countries in Asia and Africa, if not Latin America, the central historical event is decolonization. Anticolonialism and nationalism are widely accepted values endorsed by virtually all groups in the Third World. Dependency arguments are widely diffused. A Third World leader who opts for enhancing support through international behavior will reject existing rules, norms, and institutions. The most vigorous support for Third World demands for regime transformation has come from countries where such policies contributed to domestic political legitimacy. External policy has helped to define the internal character of the regime for its own constituency.

Conclusion

The countries of the Third World have not simply sought higher levels of resource transfer. They have wanted to restructure international regimes. The

New International Economic Order is the successor of SUNFED, and the First and Second Development Decades. It will be followed by other programs with different names but the same import—control, not just wealth. The NIEO, and its antecedents and probable successors, cannot be understood through analogies to reform efforts within national polities such as the labor union, consumer, welfare, and civil rights movements in the United States. These were movements based upon shared norms; the South rejects the liberal norms of the American-created postwar system. They were movements content to share power within existing structures; the South wants effective control over new structures.

The demands of the South are a function of the profound international and domestic weakness of most Third World states. These demands may temporarily abate but they will not disappear. Since most states of the South cannot hope to garner the national resource capabilities needed to assert effective control in the international system, they will continue to press for international institutions and norms that can offer them some control over the international environment. In the pursuit of this goal, they will enjoy some success by taking advantage of institutional structures that were created by the powerful to serve their own purposes. In this, and other ways, the power of hegemonic states is dissipated by the very structures they have created.

Index